The Engineer's Error Coding Handbook

The Engineer's Error
Coding Handbook

A.D. Houghton

Department of Electronic and Electrical Engineering
The University of Sheffield
Sheffield
UK

CHAPMAN & HALL

London · Weinheim · New York · Tokyo · Melbourne · Madras

Published by Chapman & Hall, 2–6 Boundary Row, London SE1 8HN, UK

Chapman & Hall, 2–6 Boundary Row, London SE1 8HN, UK

Chapman & Hall GmbH, Pappelallee 3, 69469 Weinheim, Germany

Chapman & Hall USA, 115 Fifth Avenue, New York, NY 10003, USA

Chapman & Hall Japan, ITP-Japan, Kyowa Building, 3F, 2-2-1 Hirakawacho, Chiyoda-ku, Tokyo 102, Japan

Chapman & Hall Australia, 102 Dodds Street, South Melbourne, Victoria 3205, Australia

Chapman & Hall India, R. Seshadri, 32 Second Main Road, CIT East, Madras 600 035, India

First edition 1997

© 1997 A.D. Houghton

Printed in Great Britain at T.J. Press (Padstow) Ltd., Padstow, Cornwall

ISBN 0 412 79070 X

A catalogue record for this book is available from the British Library

∞ Printed on permanent acid-free text paper, manufactured in accordance with ANSI/NISO Z39.48-1992 and ANSI/NISO Z39.48-1984 (Permanence of Paper).

Contents

Dedicated to my parents

Hugh and Hazel Houghton

Preface

Error coding is a fascinating subject as much, if not more so, as it is an indispensable part of modern engineering systems. Unfortunately, in a bid to remain general and to create a solid foundation upon which to build, many books on this subject are out of the reach of those with more engineering-based, or non-mathematical backgrounds. This is a pity because in many cases the maths is tractable with few and simple rules. If we are content to believe that it works, and let others worry about the deeper mysteries of how or why, then with a little practice the design and implementation of practical error coding systems becomes straightforward.

In this text I have attempted to reveal the useful kernel of the subject, removing the shell of terms and proofs that usually surrounds it. Being somewhat empirical in nature (an empiricist), and occasionally heard to quote the adage, 'if it works twice it's a law', my explanations take this form. For many, including myself, abstract ideas are often better grasped by practical illustration than from yards of theory.

The book is aimed at communications engineers who do not have the time to acquire an appropriate background necessary to tackle more traditional texts, and for engineering students whose maths is geared towards their own subjects. The book will be of most interest to those studying communications, electronics, signal processing and computer systems. My hope is that the book provides both a useful source for those wishing to build electronic systems, and a springboard for those who want to go on to study the subject in more depth.

WHAT YOU WILL FIND

In keeping with my aim of minimizing the maths, most ideas are conveyed by the use of simple illustrative examples so as not to overwhelm you with copious proofs, complex terms and generalized theories. You will find, therefore, the information required to do the job and a modest description (where appropriate) of what's going on and where the limits are. For interest's sake, some background is included beyond that which is absolutely necessary for the benefit of those who wish to probe a little deeper.

To build up the concepts of error detection we first examine the very simple case of parity checking, developing this into the cyclic redundancy check. For error correction, Hamming codes are first considered and their ethos extended into Reed–Solomon coding. There is no attempt here to catalogue the large array of different error control methods which are often variations on a theme. A cursory look at convolutional coding is also included, however, by virtue of the fact that it represents probably the other main strategy to the block codes given here.

The appendices include Pascal source code to allow you to repeat the experiments included in the text, and libraries of functions to allow you to perform experiments of your own.

Last, it would be most inappropriate to continue without mentioning two key authors who have between them made this subject accessible. First, John Watkinson who through his excellent books has managed to demonstrate that the subject is in fact quite simple, and second, Peter Sweeney who provided enough information to show how the theory can be usefully extended to more complex systems. I should also like to take this opportunity to thank Mr David Hatter, commissioning editor for Chapman and Hall, for his commitment to this project.

INTERNET PROVISION OF SOURCE CODE

Code for many of the examples used in the text (Appendices B to F) is available free of charge on the Internet, and this facility will be updated periodically with new ideas and suggestions. This allows you not only to repeat the experiments included, but to generate coded solutions to real problems.

Internet address: http://www.shef.ac.uk/~eee/esg/staff/adh_prof.html

Glossary of terms

Bit Error Rate (BER) is a measure of the average likelihood that a bit will be in error. A channel with a BER of 10^{-6}, for example, means that on average one bit in 1 000 000 will be received in error.

Block Codes are codes that require data being compiled into blocks before coding or decoding can take place.

Block Interleaving is a technique for packaging data so as to decorrelate large burst errors, spreading their effect over many individually coded data blocks. This means that only modest codes are required to correct potentially huge consecutive errors. Interleaving is of no value for random errors.

Burst Errors are bunched groups of error bits often caused by electrical impulses near a channel. The length b of a burst error is defined as the number of bits between (which may be correct) and including two error bits, which must be preceded by at least b error-free bits.

Coding Gain is a measure of an apparent signal-to-noise ratio improvement in a channel after error correction coding has been added. In some cases, coding gain may be negative. When this happens, the improvement accrued by adding error coding has been more than offset by the channel bandwidth required to convey the code's redundancy.

Coding Rate is the ratio of data bits to message bits for a particular arrangement. For example, a convolutional coder which outputs three (message) bits for every two (data) bits input has a coding rate of $2/3$.

Constellation is a term used to describe the pattern of possible amplitude and phase values that a phase amplitude modulated signal can have.

Convolutional Coding is the alternative to block coding, where encoding and decoding can take place on a continuous data bitstream.

CRC. See Cyclic Redundancy Check.

Cross-Interleaving is a special case of block-interleaving, used extensively in digital audio applications. Cross-interleaving has a better error-spreading capability than simple block-interleaving.

Cyclic Redundancy Check (CRC). This is an error check calculated by dividing a message by a generator polynomial (GP) and taking the remainder. The remainder (the CRC) is appended to the message such that it becomes exactly divisible by the GP at the receiver. If no remainder is found at the receiver when the message is divided by the GP then the message is deemed to be error-free.

d_{min} is the minimum Hamming distance between any valid encoded messages. d_{min} is one in an unencoded message and must be changed to $2t + 1$ where t errors are to be corrected.

Fourier Transform. This is an algorithm used to convert time domain message representations into an orthogonal frequency domain form.

Frequency Domain is one possible representation of a message. Time domain messages are converted to the frequency domain by means of a Fourier transform. The time and frequency domains are orthogonal. For the purpose of some coding strategies, messages start life in the frequency domain but are transmitted in the time domain.

Hamming Codes are messages encoded and decoded on the basis of parity check matrices. Hamming coding introduces a d_{min} of three between valid messages.

Hamming Distance is a measure of the space between two messages. It is calculated by counting how many bits must be changed to turn one message into the other.

Inverse Fourier Transform. This is an algorithm used to convert data in a frequency domain form into a time domain representation.
PAM. See Phase Amplitude Modulation.

Parity Bits are added to a message to ensure that the weight of the message bits, or (in the case of Hamming codes) combinations of certain bits are odd or even.

Phase Amplitude Modulation (PAM) is a scheme whereby a carrier is modulated in both amplitude and phase by one out of m discrete choices which make up the PAM constellation. Each signal thus conveys $\text{Log}_2(m)$ bits.

Random Errors is the name given to bit errors where the distribution of the errors is Gaussian.

Reed–Solomon Coding is a particular type of block coding highly suited to binary data sets.

RS Coding. See Reed–Solomon Coding.

TCM. See Trellis Coded Modulation.

Time Domain. This is the usual representation of a message where the data are directly viewable. If a message exists (hidden) in a frequency domain format, then an inverse Fourier transform is required to convert the message to the time domain for subsequent use.

Trellis Coding. See Convolutional Coding.

Trellis Coded Modulation (TCM) is a system which superimposes a trellis encoded data stream onto a phase amplitude modulation scheme.

Weight refers to the number of bits in a message that are set to one.

Viterbi Algorithm. This is an algorithm used to decode convolutionally coded messages. The algorithm minimizes the amount of work required to find the correct (or closest) valid data path through a trellis, based on a received data stream.

Part One

Error detection

1

Introduction

1.1 REDUNDANCY

The most important concept when dealing with either detection or correction of errors, is that of **redundancy**. When storing or transmitting data, if error control is required then redundancy must be added to the data which means that extra 'bits' (because they do not represent information as such) must be stored or transmitted along with the data. In the event of no errors these extra bits are literally redundant. If we have an eight-bit message and add an extra error control bit, then more of the communications bandwidth or disk space has been used, but we still have only eight bits of information.

Increasing the amount of data may seem anomalous in an era when data compression is so important. By virtue of their statistical properties, most data are already a combination of **information** and redundancy. Information describes the minimum amount of data that can completely convey a message while redundancy is the difference between the message and its information. This is easily understood by thinking of a digital pict're of a blank sheet of white paper. If the picture is 256×256 pixels, each represented by 8 bits, then in its raw form the picture requires 65 536 bytes of storage space or transmission bandwidth. The picture could, however, be more readily described as 65 536 pixels of intensity 255 (white). This would require 3 bytes. The information is 3 bytes while the redundancy is 65 533 bytes. Another example would be transmitting the phrase 'The cat sat on the mat'. This could be reduced to 'The cat _ _ _ mat'. The majority of people familiar with English would be able to fill in the blanks.

The problem with natural redundancy is that it depends on the type of data and it may be quite unstructured. When adding redundancy for error control, it has a very definite relationship to the data. The simplest example would be to transmit a message twice. The redundancy is equal to the message size. If at the receiver the two messages are not the same then clearly an error has occurred. This is error detection because there is no way of knowing which of the two messages is the correct one. To perform error correction we might transmit the message three times. If two out of the three messages are the same then it can be assumed that these represent the original (correct) message. Redundancy is now twice the message size and both of these examples represent very inefficient systems. The art of error control coding is arranging the redundancy so that it provides us with an efficient means of detecting and correcting errors.

1.2 ORTHOGONALITY

The term orthogonality is sometimes associated with error control coding and is basically a measure of the 'space' that exists between uncorrupted, error control encoded messages. If a message has, say, k bits and r bits of redundancy, then there are 2^{k+r} possible data patterns that can exist, but only 2^k are valid messages. Considering the simple case of adding one parity bit to each data byte, then 256 out of a possible 512 nine-bit values will be valid. Redundancy is added to make these valid messages look as different from each other as possible and this difference is loosely described as their orthogonality.

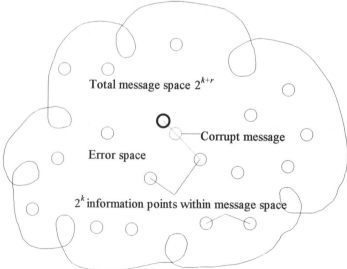

Fig. 1.1 Valid data points within the total message space.

Figure 1.1 illustrates valid messages sitting within the total message space. As messages are increasingly corrupted, (dotted) they move from valid message positions into the error space around them. If the capacity of the error coding to correct the message has not been exceeded, then the nearest valid position (bold) to the received message will be the correct message.

The reason that orthogonality is important can be illustrated by considering the simple case of the single parity bit. With a nine-bit word, eight data bits and one even parity bit, a single-bit error changes the word from one of the 256 valid messages to one of the 256 invalid messages. We have detected the error but how about correction? The example below shows a data byte plus parity.

Original data plus parity   0 1 1 1 0 1 0 1   1
After corruption   0 1 0 1 0 1 0 1   1

Unfortunately there are nine ways in which single bits could be modified in the message to correct the parity, but only one of them represents the corrected data. The **Hamming distance** is a measure of the difference between codes and is calculated by counting how many bits have to be changed to get from one code to another. A single-bit error gives rise to a Hamming distance of one between the true message and the corrupted message. In the previous example there are nine valid messages which have a Hamming distance of one from the corrupted message and there is no way of knowing which represents the correct one. This introduces another fundamental error control coding concept. If we wish to correct t bit errors, then we must ensure that there is a minimum Hamming distance (usually called d_{min}) between valid codes of $2t + 1$. In other words, if we could ensure that at least three bits would have to be changed to make one valid message look like another valid message, then a single-bit error would result in an invalid message whose Hamming distance is one from the correct message, but two or more from any other valid message.

1.3 DATA PACKAGING

As will be seen later on, there are several ways in which we can package data that will enhance the effectiveness of any error control coding strategy. More fundamental, however, is the following question. We have 100 bytes of data and wish to protect them from up to ten errors. Is it better to split the data into ten blocks of ten bytes and give each block the capacity to correct one error, or keep the data as a whole and protect against ten errors? As with many such questions the answer depends on the application. Ideally the latter solution would be chosen where the distribution of the errors is not an issue. If the data are split into blocks of ten and two errors fall within one block, then the message is irreparable. Conversely, calculation of the error control codes for the first case is trivial when compared with the latter.

1.4 SUMMARY

To summarize so far, error control coding adds redundancy to a message in such a way as to be able to identify whether or not the message has been corrupted by errors. If a correction strategy has been employed then this redundancy also allows location and correction of the errors. By adding this kind of error control coding we can introduce a **coding gain** into the message. This is equivalent to a signal-to-noise ratio improvement in a system, be it a transmission channel or a storage device. For example, a channel can give a basic SNR of say 10 dB, but after adding error control coding, an equivalent bit error rate to a channel of say 15 dB is accomplished, giving a coding gain of 5 dB. This means that increasing the size of the message through the addition of redundancy actually results in better system (or overall) performance than not doing so.

2

Parity checking

2.1 THE HORIZONTAL AND VERTICAL PARITY CHECK

You are probably already familiar with the idea of parity checking as a means of error detection. This technique has been built into many low-speed serial communications links for decades and provides a good starting place for us. In its simplest form the parity check uses one extra bit per character, where traditionally a character is a five-, seven- or eight-bit word. The two ends of the communications link must first be agreed upon the type of parity used, such as no parity, odd parity or even parity. For example, if even parity is used, then the extra **parity bit** is set so that the total number of bits in the character plus parity that are set to one (the **weight** of the message) is even. With odd parity the weight of the character plus parity is arranged to be odd. Some examples are shown in Table 2.1 below. Column E contains the even parity bits while column O shows the odd parity bits.

Table 2.1 Examples of horizontal parity

		Data						E	O
0	0	1	0	1	1	0	1	0	1
1	1	1	0	1	1	1	1	1	0
1	1	0	0	1	0	0	0	1	0
1	1	1	0	0	0	1	0	0	1
0	0	<u>0</u>	0	0	0	<u>1</u>	1	0	1

There are several things to note about a simple parity check of this nature. First and most obvious is the fact that no even bit errors (multiples of two bits) within a single character will be detected. This is unfortunate since burst errors such as are created by spikes on power supplies for example, typically comprise a few correct bits in between two error bits. The last row in Table 2.1 contains a five-bit burst error, where the third and seventh bits are in error. At the receiver these errors will pass undetected. This effect can be partly offset by including a **vertical** parity check with a block of characters. This is not always practical since communication systems that use parity are often asynchronous, making compilation of characters into a block impossible as there are arbitrary gaps between characters. Choosing even parity, Table 2.2 shows the previous data with the addition of such a vertical (even) parity check.

Table 2.2 Examples of vertical parity

Data								E
0	0	1	0	1	1	0	1	0
1	1	1	0	1	1	1	1	1
1	1	0	0	1	0	0	0	1
1	1	1	0	0	0	1	0	0
0	0	$\underline{0}$	0	0	0	$\underline{1}$	1	0
1	1	$\underline{0}$	0	1	0	$\underline{0}$	1	←Received parity
1	1	$\underline{1}$	0	1	0	$\underline{1}$	1	←Expected parity

In this case a difference between the received and expected vertical parity alerts the receiver to the presence of errors not detected by the horizontal parity checks. Even so, strategically positioned multiples of four error bits can still go unnoticed by both horizontal and vertical checks.

A second point to note about simple parity checking is that for a rather poor return in terms of detection ability, a large amount of extra bandwidth is needed. If only horizontal parity is used with eight-bit data then only 89% of the available bandwidth carries data, or 11% of the message is redundant.

2.2 GENERATING THE PARITY BIT

Parity checking is a standard feature of all universal asynchronous receiver transmitters (UARTs). These are devices which provide a convenient interface between the parallel byte or word structure of computer systems and the (mostly) serial world of communications. The UART serializes and de-serializes data as it passes between the computer and communication network (a piece of wire usually) and takes care of the character framework which has to surround data words transmitted in an asynchronous fashion. In the event of a detected transmission error the UART also has to signal this fact to the computer. While communications is not the subject in hand, generation and subsequent checking of the parity bit is a matter of interest.

To calculate the parity bit we could add up all the 1s in the word, and add a parity bit of 1 or 0 to make the final count odd or even depending on the parity selected. This would be a little slow, however, and is also unnecessary. The eXclusive OR (XOR) gate actually does this operation in parallel. A typical XOR gate has two inputs and one output. The output will be a logic 1 if the sum of the inputs is odd, or zero otherwise. Put simply, if only one of the inputs is one, the output is one. A single XOR gate will thus generate even parity for a two-bit word as shown in Fig. 2.1 below.

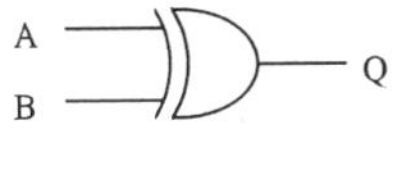

A	B	Q	P
0	0	0	0
0	1	1	1
1	0	1	1
1	1	0	0

Fig. 2.1 The eXclusive OR function.

Figure 2.1 shows the output Q which result from the two inputs A and B. The P column corresponds to the parity that would be required to make the two-bit input word (a dibit) have even parity. The XOR function can be extended to eight bits in the following ways:

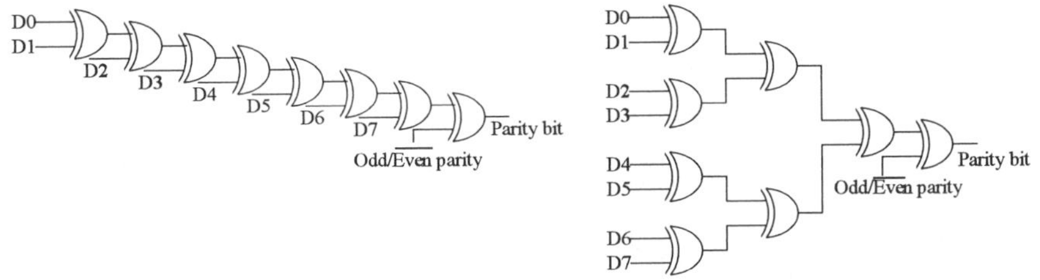

Fig. 2.2 Extending a two-input XOR gate to eight bits.

Of the two approaches shown in Fig. 2.2 for extending the XOR function, we would probably choose the right-hand side since this has half the signal propagation depth of the left-hand implementation. Although it is not of great concern here, the left-hand circuit will also have a significant ripple on the output each time the data changes. This is due to the different propagation delays that each bit experiences (apart from D0 and D1) between input and output. The extra odd/even parity input allows us to select whether the output represents an odd (1) or even (0) parity bit.

Figure 2.3 shows a typical implementation of parity within the transmitter side of a UART. A parallel data word (or byte in this case) comes from the microprocessor into a parallel transmit data register. The bits are summed along with a parity-type (odd or even) bit to produce the parity bit. When the transmit shift register (parallel in/serial out) is empty, the data word, parity bit and stop and start character framework bits (necessary in asynchronous communication) are loaded into the transmit shift register and shifted bit by bit to the outside world. The receive side of the UART looks very similar except that data bits are injected in the opposite end of the shift register (this time serial in/parallel out), and are parallel loaded into a receive data register when the complete character has arrived. At this point, the received parity bit will be compared with the recalculated parity and, if necessary, a parity error will be flagged.

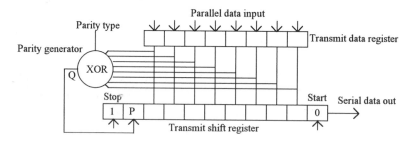

Fig. 2.3 UART organization, combining the parity bit into a serial transmission.

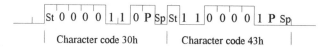

Fig. 2.4 Characters '0C' after packaging into a serial format with even parity.

Figure 2.4 shows how the voltages on the line would vary when the characters '0C' (ASCII 30h, 43h) are transmitted using seven-bit data words. Ten bits are used to transmit seven data bits, parity carrying a 10% overhead. Notice that a negative voltage is used to convey the logic state 1.

Calculation of the vertical parity check requires memory. The circuit in Fig. 2.5 shows the arrangement for a single column of bits. Only eight such circuits will be required for byte-wide data since there is no point in performing a vertical check on the horizontal parity bit as there is no guarantee that both horizontal *and* vertical parity for a random message will be satisfied.

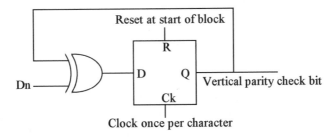

Fig. 2.5 One bit of the vertical parity checker.

At the transmitter the vertical parity checker is cleared at the start of a vertical block then clocked once per character. At the end of the vertical block the result in the registers follows the last data character. At the receiver the registers are cleared at the start of each new vertical block and clocked for each received character (including the final checksum). In the event of no errors the result in the registers will be zero after the final checksum is clocked in.

2.3 DISCUSSION

At this point you should understand how to calculate an odd or even parity bit. You should also be aware of the limitations of parity checking as a means of error detection, and how these can be overcome in some cases by the inclusion of a vertical parity check. By considering these factors and the bandwidth required by parity (10% in the previous example), you should be able to form a good idea of where parity will and will not be a useful addition to a communications or storage system. Last, you should be able to design a hardware parity checker for both horizontal and vertical cases.

3

Cyclic redundancy checking

While parity checks are useful for low data rates and asynchronous messages where large gaps in between successive bytes make compilation of data into block impossible, for more general protection of data a much more robust error detection scheme is necessary. The cyclic redundancy check or CRC is one such mechanism and it requires that the data are divided into blocks, hence it is termed a **block code**. As with many block codes the CRC is based on **finite field theory** which need not concern us greatly, except that there are a few ground rules which must be understood.

3.1 GROUND RULES AND TERMS

The first ground rule is that all operations are executed **modulo-2**. This is good news because it means that most operations boil down to eXclusive ORing, making physical implementation trivial. As a result, we do not need to be concerned with negative numbers since $1 + 1 = 0$, so $1 = -1$. The second ground rule is that all operations are bounded by the finite field in which we are operating. In short, this means that any operation that you do on the data produces a result which is contained within a finite set of numbers. Last, the finite fields that we will be using are based on numbers called **primitive polynomials** or more generally **generator polynomials**. Through examples it is my intention to try and remove any mystique that may surround such terms, bringing them down to simple bare bits. As with any branch of mathematics there are great depths that can be plumbed, but you won't find them here.

There are a variety of terms which surround the maths that is involved and these can cause a little confusion if their meaning is not clear. For reasons of generality, numbers are often expressed in the form

$$x^4 + x^2 + 1$$

for example. Since we will only be dealing with binary, x is simply 2. The above example could be written

$$x^4 + x^2 + 1 = 2^4 + 2^2 + 1 = 10101_2 = 21_{10}$$

All these representations are interchangeable, and whichever is used rather depends on the point that is being made. For example, if we wanted to multiply two such polynomials (as they are called), say 10101 and 110010, it would be wrong to assume that the answer was 21×50 (=1050). We would have to do this operation modulo-2 as

$$(x^4 + x^2 + 1)(x^5 + x^4 + x)$$

$$= \quad x^9 + x^8 + x^5 + x^7 + x^6 + x^3 + x^5 + x^4 + x$$

$$= \quad x^9 + x^8 + x^7 + x^6 + (x^5 + x^5) + x^4 + x^3 + x$$

$$= \quad x^9 + x^8 + x^7 + x^6 + x^4 + x^3 + x$$

$$= \quad 11\ 1101\ 1010 \text{ (or 986)}$$

Using conventional maths the result would have been

$$x^9 + x^8 + x^7 + x^6 + \underline{\mathbf{2x^5}} + x^4 + x^3 + x$$

but where there is an even number of similar terms the result is zero (i.e. $1 + 1 = 0$) so the x^5 term goes since $x^5 + x^5 = (1 + 1)x^5$. The assumption is made here that the field in which we are operating is large enough to contain ten bits. Were this not the case, a polynomial describing the field is used to 'wrap' the result back into the field. We will see this in action later.

3.2 AN OVERVIEW OF THE CRC

A superficial explanation of the CRC is as follows. Suppose we have some data D, and we are using a generator polynomial GP to define the field in which we are operating. The CRC is the remainder R_1 that we get when we divide D by GP as found from (3.1):

$$D = Q_1 \times GP + R_1 \tag{3.1}$$

where Q_1 is the quotient obtained from the division. When transmitting the message, the remainder is appended. This is received as

$$D + R_1 = Q_2 \times GP + R_2 \tag{3.2}$$

Since $1 = -1$ modulo-2, (3.1) may be rearranged to (3.3) as

$$D + R_1 = Q_1 \times GP \tag{3.3}$$

and substituting (3.3) into (3.2) gives

$$Q_1 \times GP = Q_2 \times GP + R_2 \tag{3.4}$$

If there have been no errors, then $Q_1 = Q_2$ thus $R_2 = 0$. If, however, R_2 is not zero then there are transmission errors. The only step that may not be clear here is the formation of (3.3) from (3.1). This simply requires us to remember that $1 = -1$ in modulo-2. We have taken our data, divided it by a polynomial and transmitted the data plus remainder. The inevitable result of doing this is that when we divide the combined number at the receiver, we expect no remainder in the absence of errors.

3.3 THE GENERATOR POLYNOMIAL

Choice of the generator polynomial is not arbitrary, and depends on several factors. Principally, the GP defines a finite field. A finite field is a set of numbers over which all calculations are performed. This simply means that the input to any calculations and the results from them have to be numbers contained by the chosen finite field. One result of this is that the GP defines the length of the CRC that is produced. If we require a 16-bit CRC (typical in many communications and storage applications) then the GP must be one bit longer, i.e. 17 bits. Of necessity, the first and last bit of the GP are 1, and the GP itself should be a primitive polynomial or derived from one. The term primitive means that the polynomial has no **factors** (i.e. it is irreducible). Consider the following:

$$(x^2 + x + 1)(x + 1)$$

expands to

$$x^3 + x^2 + x^2 + x + x + 1$$

which simplifies to

$$x^3 + 1$$

$x^3 + 1$ is **not** primitive because it has at least two factors, $x^2 + x + 1$ and $x + 1$. A primitive polynomial thus has no factors, and some examples include:

$$x^2 + x + 1 = 0,$$

$$x^3 + x + 1 = 0, x^3 + x^2 + 1 = 0,$$

$$x^4 + x + 1 = 0, x^4 + x^3 + 1 = 0,$$

$$x^5 + x^2 + 1 = 0, x^5 + x^3 + 1 = 0,$$
$$x^5 + x^3 + x^2 + x + 1 = 0,$$
$$x^5 + x^4 + x^2 + x + 1 = 0 \ldots$$

3.4 FINITE FIELDS

It is not essential to understand why the generator polynomial should be primitive, but it does provide a little more insight into the whole process and is worth a few lines. The generator polynomial defines a finite field over which all the operations take place, and the finite field can be calculated in the following way.

Taking a small field defined GF(2^3) (after Galois, originator of much of the field theory), we will use the polynomial $x^3 + x + 1 = 0$. The term GF(2^3) means Galois field over 2^3 elements. Some texts simply call this GF(8). The polynomial is four bits long so all members of the field (like the CRC) will be three bits long. To calculate the field we start off with a primitive element called α which in this case is 2 or 010 (x). Each successive member of the field is represented by an increasing power of α; α is called the primitive root because its powers represent all the non-zero members of the field. Also, for any finite field GF(2^m), $\alpha^{2^m-1} = 1$. While we won't be using the results of this computation yet, the point of interest here is that the finite field created by the GP is a (maximal) sequence of all the non-zero elements that can be described by three bits. This sequence may well already be familiar to you, but in the guise of a pseudo-random sequence. These elements obey the normal laws of maths insofar as they can be added, multiplied or divided to yield another member of the field. It should be noted that 0 is also a member of the field.

Table 3.1 GF(2^3)

α	Calculation			Numeric value
$\alpha^1 = x$	$= x$			$= 010\ (2)$
$\alpha^2 = x.x$	$= x^2$			$= 100\ (4)$
$\alpha^3 = x.x.x$	$= x^3$			

there is no term x^3, but from the primitive polynomial $x^3 + x + 1 = 0$, we can see that $x^3 = x + 1$, remembering also that $1 = -1$. Using this, we can 'fold' x^3 back into the bottom 3 bits.

α	Calculation			Numeric value
$\alpha^3 = x^3$	$= x + 1$			$= 011\ (3)$
$\alpha^4 = \alpha.\alpha^3$	$= x.(x + 1)$	$= x^2 + x$		$= 110\ (6)$
$\alpha^5 = \alpha.\alpha^4$	$= x.(x^2 + x)$	$= x^3 + x^2$	$= (x + 1) + x^2$	$= 111\ (7)$
$\alpha^6 = \alpha^2.\alpha^4$	$= x^2.(x^2 + x)$	$= x.(x + 1) + (x + 1)$	$= x^2 + 1$	$= 101\ (5)$
$\alpha^7 = \alpha.\alpha^6$	$= x.(x^2 + 1)$	$= x^3 + x^2$	$= (x + 1) + x$	$= 001\ (1)$
$\alpha^8 = \alpha.\alpha^7$	$= \alpha.1$	$= \alpha$		

A reducible or non-primitive polynomial does not produce a maximal sequence. Such a polynomial is $x^3 + x^2 + x + 1 = 0$ which has three equal factors, $(x + 1)$. Table 3.2 shows the sequence generated when using a non-primitive polynomial.

While this still does not explain the relevance of primitive polynomials to cyclic redundancy checking it does go some way towards it. To understand more clearly what effect this has we must examine the error detection capabilities of the CRC. As will be seen in some later examples, no errors equal to the GP will be detected by the GP. Superficially, if the GP divides exactly into the error, then it will yield no remainder. If, however, the GP has factors then errors related to the factors may (although not necessarily) also pass undetected, but if the GP is primitive then no errors smaller than the GP will go undetected. In short, if the GP is primitive then **all** errors smaller than the GP will be detected, yielding a well-defined performance.

Table 3.2 Generating a sequence using a non-primitive polynomial

α	Calculation			Numeric value
$\alpha^1 = x$				$= 010\ (2)$
$\alpha^2 = x.x$	$= x^2$			$= 100\ (4)$
$\alpha^3 = x.x.x$	$= x^3$			

this time, $x^3 = x^2 + x + 1$ so

$\alpha^3 = x^3$	$= x^2 + x + 1$			$= 111\ (7)$
$\alpha^4 = \alpha.\alpha^3$	$= x.(x^2 + x + 1)$	$= x^3 + x^2 + x$	$= 1$	$= 001\ (1)$
$\alpha^5 = \alpha.\alpha^4$	$= x.(1)$	$= x$	$= \alpha^1$	$= 010\ (2)$

and the cycle repeats. This time, we have not generated all non-zero elements with three bits. In fact, there are three cycles, corresponding to the three factors of $x^3 + x^2 + x + 1 = 0$. Exactly which one we generate depends on the starting conditions. The above cycle contains 2, 4, 7, 1. If we start with a value not represented here, say 3, then

$\alpha^?$	$= x + 1$			$= 011\ (3)$
$\alpha.\alpha^?$	$= x.(x + 1)$	$= x^2 + x$		$= 110\ (6)$
$\alpha.\alpha^{?+1}$	$= x.(x^2 + x)$	$= x^3 + x^2$	$= x + 1$	$= 011\ (3)$

thus we have generated a smaller cycle containing 3 and 6. If we start with the one remaining unused number, 5, then

$\alpha^?$	$= x^2 + 1$			$= 101\ (5)$
$\alpha.\alpha^?$	$= x.(x^2 + 1)$	$= x^3 + x$	$= x^2 + 1$	$= 101\ (5)$

3.5 LONGHAND CALCULATION OF THE CRC

Calculation of the CRC requires a long division, although this is done modulo-2 and so is relatively straightforward. A few variations from normal long division must be observed. We will use the primitive polynomial $x^4 + x^3 + 1 = 0$, to give a four-bit CRC in the following longhand example of Table 3.3.

Table 3.3 Longhand calculation of the CRC

```
        GP                  Quotient                      CRC

                    1  1  0  1  0  0  1  0
  1  1  0  0  1 | 1  0  1  1  0  1  1  0      0  0  0  0
                  1  1  0  0  1  ⇓
  XOR ⇒          0  1  1  1  1  1
                    1  1  0  0  1  ⇓  ⇓
      XOR ⇒        0  0  1  1  0  1  0
                       1  1  0  0  1      ⇓  ⇓  ⇓
          XOR ⇒       0  0  0  1  1      0  0  0
                             1  1      0  0  1  ⇓
              XOR ⇒         0  0      0  0  1  0
                  Remainder (CRC) =    0  0  1  0
```

The steps involved in calculating the CRC are as follows:

1. append as many zeros to the data (bold) as there will be CRC bits (one fewer than GP)
2. line up the MSB of the GP (underlined) with the first 1 (italic) in the data and XOR ($\Rightarrow$)
3. line up the MSB of the GP with the next 1 (italic), bring down data bits and zeros ($\Downarrow$) as needed
4. XOR and go back to step 3 until there are no 1s left under the data (strike through)
5. the remainder under the appended zeros (bold) is the CRC.

In this case, that transmitted message is 101101100010, comprising eight data and four check bits. The quotient is discarded serving only as a record of where the eXORing took place. At the receiver an almost identical process takes place. The only difference is that the CRC is used in the division rather than appending four zeros. This is shown in Table 3.4.

Table 3.4 Checking for errors at the receiver

```
        GP                    Quotient                 CRC

                    1  1  0  1  0  0  1  0
    1  1  0  0  1 | 1  0  1  1  0  1  1  0      0  0  1  0
                    1  1  0  0  1  ⇓
                    0  1  1  1  1  1
                       1  1  0  0  1  ⇓  ⇓
                       0  0  1  1  0  1  0
                          1  1  0  0  1      ⇓  ⇓  ⇓
                          0  0  0  1  1      0  0  1
                                   1  1      0  0  1  ⇓
                                   0  0      0  0  0  0
                        Remainder = 0        0  0  0  0
```

In the absence of errors the remainder is zero, but if a detectable error has occurred, then it will be non-zero. An example is given in Table 3.5 where the bits shown bold have been inverted.

Table 3.5 Decoding a message with errors

```
        GP                    Quotient                 CRC

                    1  1  0  1  0  1  0  1
    1  1  0  0  1 | 1  0  1  1  0  0  1  0      0  1  1  0
                    1  1  0  0  1  ⇓
                    0  1  1  1  1  0
                       1  1  0  0  1  ⇓  ⇓
                       0  0  1  1  1  1  0
                          1  1  0  0  1      ⇓  ⇓
                          0  0  1  1  1      0  1
                                1  1  0      0  1  ⇓  ⇓
                                0  0  1      0  0  1  0
                                      1      1  0  0  1
                                      0      1  0  1  1
                          Remainder ≠ 0      1  0  1  1
```

If, however, we arrange for the error pattern to be equal to the GP, then the error will not be detected. To do this we simply line up the GP with the data + CRC at some arbitrary point, and XOR. Table 3.6 gives an example of this. Again, the errors are shown in bold.

Table 3.6 Non-detection of certain error patterns

```
           GP                    Quotient                CRC
                        1  1  0  1  0  1  1  0
   1  1  0  0  1 | 1  0  1  1  0  0  0  0      0  1  1  0
                   1  1  0  0  1  ⇓
                   0  1  1  1  1  0
                      1  1  0  0  1  ⇓  ⇓
                      0  0  1  1  1  0  0
                         1  1  0  0  1     ⇓  ⇓
                         0  0  1  0  1     0  1
                               1  1  0     0  1  ⇓
                               0  1  1     0  0  1
                                  1  1     0  0  1  ⇓
                                  0  0     0  0  0  0
                        Remainder = 0      0  0  0  0
```

By adding errors equal to the GP we have effectively created another valid data + CRC pattern. That this can happen is an inevitable result if we consider the following. In these examples we have used eight-bit data which means that there are 256 valid data bit patterns. Since the CRC is only four bits, it can only have 16 different values. The 256 data values must therefore map onto only 16 possible CRCs which in turn means that 16 data patterns will share the same CRC. The corollary to this that 1 in 16 errors will go undetected. This may seem a disappointing result until you consider that in reality the CRC will usually be 16 bits long, missing only 1 in 65 536 of all errors. This gets even better when we realize that the majority of burst errors will be 16 bits or less in length, guaranteeing detection.

3.6 PERFORMANCE

Performance can be measured in a variety of ways including detection ability, bandwidth requirements and ease of implementation. In a typical application a 16-bit CRC will be used to protect, say, 2 Kbits of data. In terms of bandwidth, this means that 1 bit in 129 is used for error checking compared with 1 bit in 9 for simple parity, or 0.78% of the channel instead of 11.1%. As far as detection ability goes, we can guarantee that all burst errors smaller in length than the GP will be detected if the GP is primitive, and 2^m-1 out of 2^m errors in total where there are m bits in the CRC. For a 16-bit CRC this works out to 99.9985% of errors. This calculation comes from the rash assumption that a random error gives rise to a random result in the remainder. If the remainder is m bits then 1 in 2^m random results will be zero and so not detected.

If there are enough bits in the remainder to uniquely describe all data and CRC bit locations plus one, then a single-bit error can be corrected since each such error produces a unique remainder. In the previous examples, the CRC was four bits, so provided the data + CRC is less than 16 bits, a one-bit error correction can be implemented. The danger of such a strategy, however, is assuming that only a single-bit error has occurred. If more than one bit was in error then any attempt at correction would almost certainly introduce an extra error, but worse than this, may also convey a false sense of security in the validity of the data.

3.7 HARDWARE IMPLEMENTATION

Hardware implementation is mostly straightforward since all operations are eXclusive ORing. There are as many edge-triggered registers as there are bits in the CRC, one fewer XOR gates (2 input) as there are '1's in the GP, and a single AND gate. To construct the CRC generator, line up the GP, MSB to the left so that the bits fall in between the registers, and connect as shown in Fig. 3.1. Where there is a '0' in the GP each Q output connects directly to the adjacent D input. If there is a '1' in the GP, connection is via an XOR gate, taking the second input to the XOR gate from a feedback path.

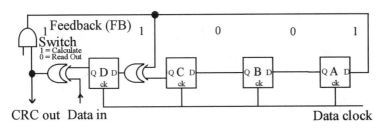

Fig. 3.1 Hardware CRC calculator.

Initially the registers are reset to zero, and the data are clocked serially into the circuit with the switch in calculate mode. When the final data bit has been clocked in (no zeros added), the CRC is actually in the registers. To clock it out without corrupting it, the switch is set to read out which disables the feedback path. Three further clocks serially output the CRC.

To verify the operation of the circuit we can use the data from the previous example, both before and after corruption by errors. Table 3.7 shows the initial calculation of the CRC. After all the data have been clocked in, the CRC resides in the four registers.

Table 3.7 Calculation of the CRC by hardware

$A = FB'$	$B = A'$	$C = B'$	$D = C' \; XOR \; FB'$	$DATA$	$FB = DATA \; XOR \; D$
0	0	0	0	1	1
1	0	0	1	0	1
1	1	0	1	1	0
0	1	1	0	1	1
1	0	1	0	0	0
0	1	0	1	1	0
0	0	1	0	1	1
1	0	0	0	0	0

The last bit of data is clocked in at this point and the AND gate switch is OPENED (=0) to prevent the feedback from destroying the CRC now in *ABCD*. The CRC clocks serially out of the registers MSB First.

0	1	0	0	0 (0)	0 (0)
0	0	1	0	0 (0)	0 (0)
0	0	0	1	0 (1)	1 (0)
0	0	0	0	0 (0)	0 (0)

Table 3.7 also shows, in brackets, the situation at the receiver where the CRC is appended to the data stream. In the absence of errors this results in four zeros. Table 3.8 shows the situation given previously in Table 3.5 where two errors have occurred. Again, the hardware result agrees with the longhand calculations given previously. Appendix A gives a list of primitive polynomials that may be useful for those wishing to experiment further. Standard polynomials for CRC generation include

$$x^{16} + x^{12} + x^5 + 1 \qquad (\equiv 1\;0001\;0000\;0010\;0001 \text{ or } 69\;665)$$

and

$$x^{16} + x^{15} + x^2 + 1 \quad (\equiv 1\ 1000\ 0000\ 0000\ 0101 \text{ or } 98\ 309)$$

for 16 bits, and

$$x^{12} + x^{11} + x^3 + x^2 + x + 1 \ (\equiv 1\ 1000\ 0000\ 1111 \text{ or } 6159)$$

for 12 bits.

Table 3.8 Calculation of the remainder with errors

$A = FB'$	$B = A'$	$C = B'$	$D = C'\ XOR\ FB'$	$DATA$	$FB = DATA\ XOR\ D$
0	0	0	0	1	1
1	0	0	1	0	1
1	1	0	1	1	0
0	1	1	0	1	1
1	0	1	0	0	0
0	1	0	1	0	1
1	0	1	1	1	0
0	1	0	1	0	1
1	0	1	1	0	1
0	1	0	1	1	0
0	0	1	0	1	1
0	0	0	1	0	1

All the polynomials listed above have the factor $(x + 1)$, for example

$$x^{16} + x^{12} + x^5 + 1 = (x + 1)(x^{15} + x^{14} + x^{13} + x^{12} + x^4 + x^3 + x^2 + x + 1)$$

By multiplying a primitive polynomial by $(x + 1)$, the CRC length is increased by 1, but the length of the cycle remains the same. We have seen that the polynomial $x^3 + x + 1 = 0$ gives rise to a repeating sequence (Table 3.1) of seven non-zero elements. If we multiply this polynomial by $(x + 1)$ we have $x^4 + x^3 + x^2 + 1 = 0$. The element lengths and CRC are one bit longer but the cycle is still only seven elements long. Starting with $\alpha = x$, the cycle for this polynomial is given in Table 3.9.

These are called expurgated cyclic codes and if we trade one of our data bits for the extra CRC bit so that, say, a seven-bit message with four data bits as could be generated by $x^3 + x + 1 = 0$ becomes a seven-bit message with only three data bits, then the minimum Hamming distance (d_{min})

between valid codes is increased by one. The implications of this will be seen more clearly when we consider cyclic codes as a means of error correction.

Table 3.9 Seven-element sequence using $x^4 + x^3 + x^2 + 1 = 0$

Power	Binary value
α	0010
α^2	0100
α^3	1000
α^4	1101
α^5	0111
α^6	1110
α^7	0001

3.8 DISCUSSION

Where data are to be organized into blocks you will see that the CRC provides a greatly enhanced error detection capability over simple parity checking, and at a greatly reduced cost in terms of bandwidth. A 16-bit CRC appended to a typical message of, say, 2 Kbits uses about 0.8% of the channel as opposed to around 11% for parity. For infrequent and intermittent communication, however, parity will be more applicable because bandwidth is not usually an issue in this case, and compilation of data words into blocks is not possible.

You should be able to calculate a CRC on a block of data given a suitable generator polynomial, and derive a hardware generator from the polynomial bit pattern. Given a primitive polynomial you should also be able to demonstrate that it is primitive and create a sequence of non-zero elements based upon the polynomial.

Part Two

Error correction, an introduction

4

Hamming codes

We have already passed over rudimentary error correction in the form of both parity checking and the cyclic redundancy check. Table 2.2 shows an example of vertical parity checking in conjunction with horizontal parity checking. The purpose of that particular example was to illustrate that some of the shortcomings of the simple horizontal parity check could be overcome. If, however, only a single-bit error occurs then the bit which is intersected by both horizontal and vertical parity errors will be the bit in error, and it can be corrected.

Rather than look at the very simple case of intersecting horizontal and vertical parity checks as a mechanism for error correction, we'll look at the slightly more general case of Hamming codes which are based on parity checks. Hamming codes represent the first serious attempt at providing a mechanism for error correction and they work by introducing a minimum Hamming distance of $2t + 1$ between valid codes. Such codes are often referred to as (n, k) codes, where n is the total number of bits in each codeword, and k is the number of data bits in the codeword. Clearly $n - k$ bits are, therefore, redundancy associated with increasing the Hamming distance between the codewords.

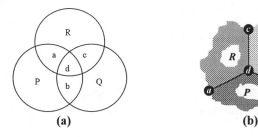

(a)   **(b)**

Fig. 4.1 Calculating a (7, 4) code.

If we consider for example a simple (7, 4) code, it can be constructed as follows. Figure 4.1a shows a Venn diagram containing four data bits (k), a, *b*, *c* and *d* and three parity bits (*n* − *k*), *P*, *Q* and *R*. For any given data, *P*, *Q* and *R* are set to 0 or 1 in order to fix the parity in each of the circles to a known state, typically even. In principle this is exactly the same as the horizontal and vertical parity check except that the data are now arranged in a three-dimensional format and the parity is calculated in each plane.

At the receiver the parity is recalculated based on the received codeword. In the event of a single-bit error, the parity in one or more of the circles will be incorrect. By noting which circles exhibit a parity error the erroneous bit can be isolated. If, for example, the parities in the circles containing *P* and *Q* were in error, the only bit common to both circles is *b*. If only the circle containing *Q* was in parity error then *Q* would be the error bit since it alone is associated only with that circle. This example goes a long way towards explaining the use of the rather abstract term orthogonality in error coding. Figure 4.1b actually shows that data arranged in such a manner that they sit orthogonally to each other. With a little stretching of the imagination you can see how the parity bits are formed from orthogonal combinations of data bits. In serious error coding, of course, the orthogonal relationships are far less obvious than here where we can comfortably visualize the problem in three-dimensional space, but the principle is the same. If you can grasp this, then you're nearly there.

From Fig. 4.1 we can construct the following orthogonal (thrown in to make the point) equations to describe the generation of *P*, *Q* and *R* from *a*, *b*, *c* and *d*.

$$
\begin{aligned}
P &= a + b + d \\
Q &= b + c + d \\
R &= a + c + d
\end{aligned}
$$

These in turn give rise to the codewords in Table 4.1 below. A careful examination of the codewords reveals that at least three bits must be changed to get from one codeword to another codeword. The code, therefore, satisfies the requirement that the minimum Hamming distance must be $2t + 1$ between codes, and *t* is one for a single-bit error correction. A useful property of the codewords is that they are **linear**. This simply means that we can combine them (by eXORing), and we will always produce another valid codeword. By storing only the codewords for one, two, four and eight, we can produce all the others. For example, if we take the code for one (0001 101) and add (XOR) to it the code for eight (1000 111) we get the code for nine (1001 010). Using Table 4.1 you can try other combinations to verify this.

Table 4.1 Codewords from the (7, 4) code

Code	0 1 2 3	4 5 6 7	8 9 A B	C D E F
d	0 0 0 0	0 0 0 0	1 1 1 1	1 1 1 1
c	0 0 0 0	1 1 1 1	0 0 0 0	1 1 1 1
b	0 0 1 1	0 0 1 1	0 0 1 1	0 0 1 1
a	0 1 0 1	0 1 0 1	0 1 0 1	0 1 0 1
P	0 1 1 0	0 1 1 0	1 0 0 1	1 0 0 1
Q	0 0 1 1	1 1 0 0	1 1 0 0	0 0 1 1
R	0 1 0 1	1 0 1 0	1 0 1 0	0 1 0 1

4.1 CORRECTING A SINGLE BIT

Error detection and correction can be performed by means of a parity check matrix. This matrix is formed by rearranging the parity generation equations above, and transposing as in Table 4.2 below.

Table 4.2 Generating a parity check matrix

d +		b +	a +	P			= 0
d +	c +	b +			Q		= 0
d +	c +		a +			R	= 0

1	0	1	1	1	0	0
1	1	1	0	0	1	0
1	1	0	1	0	0	1

1	1	1
0	1	1
1	1	0
1	0	1
1	0	0
0	1	0
0	0	1

In order to check for (and correct) errors, the received (7, 4) codeword is multiplied by the parity check matrix to generate a three-bit **syndrome**. A result of zero indicates that the received codeword is valid. Using the codeword for the number nine, we find the following:

$$[1\,0\,0\,1\,0\,1\,0] \times \begin{bmatrix} 1 & 1 & 1 \\ 0 & 1 & 1 \\ 1 & 1 & 0 \\ 1 & 0 & 1 \\ 1 & 0 & 0 \\ 0 & 1 & 0 \\ 0 & 0 & 1 \end{bmatrix} = \begin{bmatrix} (1.1 \oplus 0.0 \oplus 0.1 \oplus 1.1 \oplus 0.1 \oplus 1.0 \oplus 0.0) \\ (1.1 \oplus 0.1 \oplus 0.1 \oplus 1.0 \oplus 0.0 \oplus 1.1 \oplus 0.0) \\ (1.1 \oplus 0.1 \oplus 0.0 \oplus 1.1 \oplus 0.0 \oplus 1.0 \oplus 0.1) \end{bmatrix}$$

$$= [0, 0, 0]$$

the syndrome is zero. If we introduce a single-bit error into the codeword and repeat the operation then the syndrome takes on a non-zero value.

$$[1\,\mathit{1}\,0\,1\,0\,1\,0] \times \begin{bmatrix} 1 & 1 & 1 \\ 0 & 1 & 1 \\ 1 & 1 & 0 \\ 1 & 0 & 1 \\ 1 & 0 & 0 \\ 0 & 1 & 0 \\ 0 & 0 & 1 \end{bmatrix} = \begin{bmatrix} (1.1 \oplus \mathit{1}.0 \oplus 0.1 \oplus 1.1 \oplus 0.1 \oplus 1.0 \oplus 0.0) \\ (1.1 \oplus \mathit{1}.1 \oplus 0.1 \oplus 1.0 \oplus 0.0 \oplus 1.1 \oplus 0.0) \\ (1.1 \oplus \mathit{1}.1 \oplus 0.0 \oplus 1.1 \oplus 0.0 \oplus 1.0 \oplus 0.1) \end{bmatrix}$$

$$= [0, 1, 1]$$

To isolate the error the syndrome is compared to the parity check matrix. It appears on the second row down, which is the row that multiplies with the second bit in the codeword. This of course is the error bit and it must be inverted to correct the codeword.

As there are only enough parity bits (resulting in syndrome bits) to describe seven *bit* locations plus no errors, an inevitable result is that multiple-bit errors always give rise to a syndrome which cannot be distinguished from that of a single-bit error, or no error. An attempt at correction will always lead to an incorrect result in this case. In order to make the scheme more robust, a greater minimum Hamming distance must be placed between codewords. The code is defined as a **perfect** code because of this property. In other words, there exist no messages with a Hamming distance of more than *t* to a valid codeword.

The use of a parity check matrix looks impressive, but may make the calculation and subsequent checking of the Hamming codes appear rather more tricky than it actually is. As far as encoding and decoding the data word is concerned, the circuits of Fig. 4.2 will suffice.

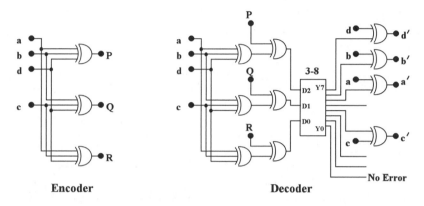

Encoder   **Decoder**

Fig. 4.2 A (7, 4) Hamming encoder and decoder.

In the encoder, the appropriate bits are summed (eXclusive-ORed) to produce the three parity bits, and this process is repeated at the decoder. The received parity bits are XORed with the recalculated versions to yield the syndrome. The syndrome feeds into a three-to-eight-line decoder (with true outputs). A zero syndrome results in the 'No Error' output (Y0) going high. A non-zero syndrome results in one of Y1 to Y7 going high which then inverts the appropriate received bit via the final XOR gates at the right. The corrected outputs are labelled d', c', b', a'. There is little point in correcting the parity bits since these are not used again unless the message is to be passed on.

4.2 EXTENDING THE MESSAGE SIZE

Extension of Hamming codes to longer messages becomes increasingly difficult to visualize. The representative Venn diagram (Fig. 4.1a) has to include 'magic tunnels' in order to provide greater dimensionality to the systems of equations that result. Implementation, however, is as trivial as for the (7, 4) code above. If we extend the message such that there are four parity bits, then the syndrome can have 16 values, enough for 15 errors plus a no-error code of 0. So the complete message is 15 bits, or a (15, 11) code. To construct the parity check matrix the data and parity bits are ordered as follows:

$$d_F, d_E, d_D, d_C, d_B, d_A, d_9, p_8, d_7, d_6, d_5, p_4, d_3, p_2, p_1$$

The bit numbering starts with 1 at the right, increasing to the left. The parity bits are placed in positions corresponding to increasing powers of 2, i.e. 1, 2, 4, 8, while data bits fill the remaining spaces.

Table 4.3 Generating the parity check equations for a (15, 11) code

	d_F	d_E	d_D	d_C	d_B	d_A	d_9	p_8	d_7	d_6	d_5	p_4	d_3	p_2	p_1
p_1	▓		▓		▓		▓		▓		▓		▓		▓
p_2	▓	▓			▓	▓			▓	▓			▓	▓	
p_4	▓	▓	▓	▓					▓	▓	▓	▓			
p_8	▓	▓	▓	▓	▓	▓	▓	▓							

Table 4.3 shows the four parity bits in the left-hand column, and the 15 message bits along the top row. The parity bits must be set or cleared so that all non-shaded elements in each row add up to zero (for even parity). From the table it can be seen that the p_1 row contains only the parity bit p_1. Likewise, the other rows contain only their own respective parity bits. This means that the outcome of any parity bit will not affect the calculation of any other parity bits. From Table 4.3 the following parity generator equations can be generated:

$$p_1 = d_F + d_D + d_B + d_9 + d_7 + d_5 + d_3$$
$$p_2 = d_F + d_E + d_B + d_A + d_7 + d_6 + d_3$$
$$p_4 = d_F + d_E + d_D + d_C + d_7 + d_6 + d_5$$
$$p_8 = d_F + d_E + d_D + d_C + d_B + d_A + d_9$$

Each parity bit is associated with one of the four message index bits. For example, p_1 is associated with all message bits whose position index within the message is odd, or has the least significant bit set, i.e. bits 1 (p_1), 3 (d_3), 5 (d_5) etc. Table 4.4 shows the calculation of the parity bits for the 11-bit data word 11011010100.

Table 4.4 Calculating the parity bits for a (15, 11) Hamming code

	1	1	0	1	1	0	1	P_8	0	1	0	P_4	0	P_2	P_1
P_1	1		0		1		1		0		0		0		1
P_2	1	1			1	0			0	1			0	0	
P_4	1	1	0	1					0	1	0	0			
P_8	1	1	0	1	1	0	1	1							

In each row the parity bit is either set or cleared to give the row weight even parity (a binary summation of 0). With the test data, this gives rise to the parity code 1001. The way that the data and parity are ordered during transmission is a matter of preference. In a bus-oriented system it will probably be more convenient to keep the data together, and keep the parity apart. In a serial communication system it makes little difference whether the parity bits are interspersed with the data, or kept separate. Keeping the parity separate in this example, the 15-bit word 110110101001001 would be transmitted. Table 4.5 shows the situation at the receiver when an error occurs in d_B.

Table 4.5 Locating a single-bit error in a (15, 11) Hamming code

	1	1	0	1	0	0	1	P_8	0	1	0	P_4	0	P_2	P_1	P_e
P_1	1		0		0		1		0		0		0		0	1
P_2	1	1			0	0			0	1			0	1		1
P_4	1	1	0	1					0	1	0	0				0
P_8	1	1	0	1	0	0	1	0								1

The right-hand column of Table 4.5 indicates a parity error. If the recalculated parity does not match the transmitted parity, then a 1 is placed in the appropriate box. In this case the syndrome due to the error is 1011. 1011 is, of course, 11_{10} or B_{16} and points to bit d_B in the message which must be inverted.

Like the (7, 4) code, this does not show if there is more than a single bit in error. We can arrange for the Hamming distance to be increased by one from three to four, however, by adding an overall parity bit to the message. When a single-bit error occurs the overall parity of the message will be wrong, and the inner (correcting) parity bits will point to the error. In the case of a double-bit error the weight of the message will still be consistent with the extra parity bit (hence no overall parity error), but the inner parity bits will be indicating that an error has occurred. We are thus alerted to an uncorrectable error.

4.3 DISCUSSION

Hamming error coding is ideal for local communications across a high-performance computer backplane. In this environment bit errors are few in a properly designed system and the cost of including such coding amounts to little more than a few extra bits on the bus. High-performance memory systems also sometimes use these sorts of codes to perform internal memory correction in critical applications. It should be noticed that by extending this process to large message sizes, parallel computation of the parity bits becomes impractical. If we have to resort to serial computation of the parity, then there are better ways of approaching error correction. Probably the main strength of Hamming codes is their immediate parallel correction ability over small messages.

You should now start to have some feel for orthogonality in the context of error coding. While this is not essential for implementing error control systems, it will help in the understanding of how they work, and give you ideas for creating your own systems. You should also be able to construct parity generator equations for Hamming codes to give single-bit error correction capability over certain message sizes and perform error correction using your parity bits. Examples of standard devices for performing this kind of error correction include the TTL devices SN74LS630 to 637.

5

Error correction using the CRC

Hamming codes work well for small messages where parallel implementation of the computation is possible, but for longer messages the CRC provides a more economic solution. A single-bit error can be corrected provided the CRC is long enough to describe every bit location within the protected part of the message, and that the CRC is based on a primitive generator polynomial. If the CRC is m bits, then the total protected message length (i.e. including the CRC itself) must be less than 2^m bits. The generator polynomial used in Chapter 3 was $x^4 + x^3 + 1$. Using the primitive element $\alpha = 2$ (or x), a finite field can be constructed in the same way as shown in Table 3.1, except in this case there will be 15 non-zero elements rather than seven owing to the larger (five-bit) primitive polynomial. This time the elements are listed in Table 5.1.

Again, a maximal sequence has been generated of all the non-zero elements of the field, but how does this help? From Table 3.2, we know that the CRC for the data 10110110 is 0010, using this GP. The complete message is thus 101101100010, and we'll define the right-hand bit as position 0, increasing as we move to the left. The position index is consistent with the power of two multipliers for each bit if the message is thought of as a binary word.

5.1 LOCATING SINGLE-BIT ERRORS

Table 5.2 shows a longhand reworking of the calculation of the remainder at the receiver, except that a single-bit error has been introduced at bit position 6, shown italic and bold. The remainder is 15 which, from Table 5.1 above, coincides with the field element α^6. Repeating the operation in Table 5.3, but this time with an error in bit position 9, we have a remainder of 5 which, from Table 5.1 above, is consistent with α^9. So from this we see that a single-bit error at position k in the message, gives rise to a remainder of α^k at the receiver. You might wonder at the case where $k = 0$, i.e. the LSB is in error. As with normal maths, α^0 is 1, or α^{15}. If the remainder is 1, then we know that $k = 0$.

Table 5.1 Generating the finite field for $x^4 + x^3 + 1 = 0$

Power	Evaluation		Decimal
α	x		2
α^2	$x.x$	x^2	4
α^3	$x.x^2$	x^3	8
α^4	$x.x^3$	$x^3 + 1$	9
α^5	$x(x^3 + 1)$	$x^3 + x + 1$	11
α^6	$x(x^3 + x + 1)$	$x^3 + x^2 + x + 1$	15
α^7	$x(x^3 + x^2 + x + 1)$	$x^2 + x + 1$	7
α^8	$x(x^2 + x + 1)$	$x^3 + x^2 + x$	14
α^9	$x(x^3 + x^2 + x)$	$x^2 + 1$	5
α^{10}	$x(x^2 + 1)$	$x^3 + x$	10
α^{11}	$x(x^3 + x)$	$x^3 + x^2 + 1$	13
α^{12}	$x(x^3 + x^2 + 1)$	$x + 1$	3
α^{13}	$x(x + 1)$	$x^2 + x$	6
α^{14}	$x(x^2 + x)$	$x^3 + x^2$	12
α^{15}	$x(x^3 + x^2)$	1	1
α	$x(1)$	x	2

Table 5.2 Locating an error at bit 6 using the CRC

```
       GP                 Quotient                  CRC

                  1 1 0 1 0 0 1 0
  1 1 0 0 1 | 1 0 1 1 0 0 1 0          0 0 1 0
              1 1 0 0 1 ⇓
              1 1 1 1 0
              1 1 0 0 1 ⇓ ⇓
                1 1 1 1 0
                1 1 0 0 1      ⇓ ⇓
                  1 1 1        0 0
                  1 1 0        0 1 ⇓ ⇓
                    1          0 1 1 0
                    1          1 0 0 1
              Remainder =      1 1 1 1
```

Table 5.3 Locating an error at bit 9 using the CRC

```
        GP                    Quotient                      CRC

                      1  1  0  1  0  0  1  0
 1  1  0  0  1 | 1  0  0  1  0  1  1  0        0  0  1  0
                1  1  0  0  1  ⇓
                1  0  1  1  1
                1  1  0  0  1  ⇓
                   1  1  1  0  1
                   1  1  0  0  1  ⇓        ⇓
                      1  0  0  0          0
                      1  1  0  0          1  ⇓
                         1  0  0          1  0
                         1  1  0          0  1  ⇓
                            1  0          1  1  1
                            1  1          0  0  1  ⇓
                               1          1  1  0  0
                               1          1  0  0  1
                    Remainder =           0  1  0  1
```

5.2 A HARDWARE ERROR LOCATOR

As far as correction is concerned we could simply use a look-up table to translate the remainder into its corresponding power of α (k), and invert bit k. This is fine for small fields but could be a little unwieldy if the purpose is to fit the error detection/correction circuits onto a single low-cost IC. If time is not a problem we can actually use the CRC circuit itself to perform the error location. You will have noticed that the CRC calculator of Fig. 3.1 is the same circuit as the checker. To add also to this its ability to correct the data is great news for the chip designer.

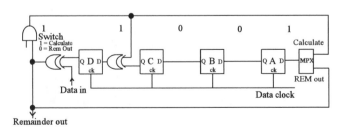

Fig. 5.1 Extending the CRC generator/checker to correction.

Figure 5.1 shows the basic CRC generator circuit with a small modification to allow correction. The input to the right-hand register is now controlled by a 2-to-1 multiplexer. During calculate, while the data bits within the message are arriving, the input is sourced as normal from the feedback path. As the CRC bits arrive at the end of the message the switch is set to REM Out, or remainder out, and the right-hand register now takes its input from the remainder output. This has the effect of placing the remainder in the registers and as soon as they are full the switch is returned to calculate mode.

As will be demonstrated somewhat later, the circuit has the effect of multiplying its contents by α. In the example of Table 5.2 the remainder was 1111 corresponding to α^6, and a bit error in position 6. If this remainder is replaced in the circuit, then after one clock the registers will contain α^7. If we count the number of clock cycles that it takes for a non-zero remainder to return to the value 1 (or α^0), then we have located the error. For α^6 we will have to clock the circuit nine times. Since we are counting forwards, when we ought to be counting backwards, this number represents $15 - k$ where 15 is the sequence length, found from $2^m - 1$ where the CRC is m bits. So in this case, $k = 15 - clocks$. If the remainder is 1, then we need not perform all 15 clocks only to produce the value 15 to subtract it from 15. To illustrate the principle, consider the second example above where the remainder is 0101 (α^9).

First, the registers are pre-loaded with the remainder 0101, shown bold in Table 5.4. (Note they appear in reverse order in the table.) Successive clocks are then applied to the circuit until the register contents are 0001. At this point, we have reached the error index, shown bold within a double box, in the $15 - Clocks$ column.

Table 5.4 Locating a single-bit error using hardware

$A = FB'$	$B = A'$	$C = B'$	$D = C'XOR FB'$	$FB = D$	$15 - Clocks$	$DCBA = 1$
1	**0**	**1**	**0**	0	15	✗
0	1	0	1	1	14	✗
1	0	1	1	1	13	✗
1	1	0	0	0	12	✗
0	1	1	0	0	11	✗
0	0	1	1	1	10	✗
1	0	0	0	0	**9**	✓

Although the maths involved won't be clear yet, it is possible to configure a circuit to multiply by α^{-1}. This means that we can count down so ending up with the correct count and avoiding the subtraction. Where there is space on an integrated circuit for this extra feature and the message length is substantially shorter than 2^m bits, this represents a faster and so attractive possibility. Later you'll find out how to generate multipliers for all powers of α over this particular field ($x^4 + x^3 + 1$), but you'll have to read on a little to find out which one to use.

5.3 MULTIPLE-BIT ERRORS

For the sake of thoroughness, we should consider the effect of more than a one-bit error on the remainder. Tables 5.2 and 5.3 show two examples of single-bit errors, but what would happen if they occurred together to form a double error? Table 5.5 shows a rework of the division with both errors present. This time the remainder is 1010 or α^{10} (from Table 5.1). Now the remainders for the two errors singly were 1111 and 0101, and if we add them, which is eXclusive ORing over finite fields, we get 1010. In other words, if we sum the remainders for the individual bit errors, we end up with the remainder for the combined errors. It is not difficult to see now (at least empirically) why errors equal to the GP will not be detected. If we sum any elements in the proportions $\alpha^0 + \alpha^3 + \alpha^4$, or multiples thereof, we always end up with zero. Recalling the previous section on Hamming distance to mind, the effect of adding three-bit errors in the proportions of the GP is to change $2t + 1$ bits (in this case) creating another valid codeword. We had already predicted that this was possible when considering the CRC as a mechanism for error detection.

Returning to the problem of two errors, we may have a means of recognizing that two errors have occurred. If the message length is less than $2^m - 1$, then there are powers of α which point to errors beyond the end of the message. Since all errors must lie within the message, it is safe to assume that in this case, more than a single-bit error has occurred. Unfortunately it is possible to add different powers of α together to yield quite small values of k. Over the field used here, errors at positions 0 and 3 would combine to create a remainder which was consistent with a single error at position 4. Clearly though, the shorter the message with respect to the CRC size, the higher the probability of detecting multiple-bit errors, and hence being able to reliably correct single-bit errors. Where the message is smaller than the cycle length of the generator polynomial, the codes are called shortened cyclic codes. The polynomial used in these examples will work for messages up to 15 bits. Shortening the message as has been done here (to 12 bits) will not decrease the minimum Hamming distance between valid codes, but may increase it.

Table 5.5 Finding the remainder for two errors

GP	Quotient	CRC
	1 1 0 1 0 0 1 0	
1 1 0 0 1 | 1 0 **0** 1 0 **0** 1 0		0 0 1 0
1 1 0 0 1 ⇓		
1 0 1 1 0		
1 1 0 0 1 ⇓		
1 1 1 1 1		
1 1 0 0 1 ⇓	⇓	
1 1 0 0	0	
1 1 0 0	1 ⇓ ⇓ ⇓	
Remainder =	1 0 1 0	

5.4 EXPURGATED CODES

Expurgated codes were mentioned in the context of error detection using the cyclic redundancy check. In the previous example, it was shown that simply shortening the message size does not guarantee that two-bit errors will be distinguishable from a single-bit error even though the probability of this increases. Using a (7, 3) code rather than a (15, 11) code with $x^4 + x^3 + 1 = 0$, errors at bit positions 0 and 3 look like a single error at bit position 4. You can verify this by adding $\alpha^0 + \alpha^3$. Multiplying the generator polynomial by $(x + 1)$ gives $x^5 + x^3 + x + 1 = 0$. This still generates a sequence of 15 non-zero elements, but produces a CRC of five bits rather than four. Table 5.6 below shows the 15 element sequence.

By trading one of the possible 11 data bits of a (15, 11) code for the extra CRC bit (expurgation) the message length stays the same but the minimum Hamming distance between codes is increased by 1. The resulting (15, 10) code thus has a d_{min} of 4 rather than 3. This means that although we cannot correct two-bit errors (for which d_{min} must be 5 or more), if two errors occur their remainder will never be mistaken for that of a single-bit error. You can check this result by examining Table 5.6. The sum of any two of the listed powers of α will never be another member of the sequence. In other words, the remainder produced by a two-bit error will result in a value not represented in the table above. The maths involved in expurgation is a little beyond the scope of this book, but it is a useful tool and costs little to implement.

Table 5.6 Cyclic codes for $(x^4 + x^3 + 1)(x + 1)$

Power	Function	Binary value	Decimal value
α	x	00010	2
α^2	x^2	00100	4
α^3	x^3	01000	8
α^4	x^4	10000	16
α^5	$x^3 + x + 1$	01011	11
α^6	$x^4 + x^2 + x$	10110	22
α^7	$x^2 + x + 1$	00111	7
α^8	$x^3 + x^2 + x$	01110	14
α^9	$x^4 + x^3 + x^2$	11100	28
α^{10}	$x^4 + x + 1$	10011	19
α^{11}	$x^3 + x^2 + 1$	01101	13
α^{12}	$x^4 + x^3 + x$	11010	26
α^{13}	$x^4 + x^3 + x^2 + x + 1$	11111	31
α^{14}	$x^4 + x^2 + 1$	10101	21
α^{15}	1	00001	1

5.5 SOFT DECISION DECODING

Traditionally a slicing circuit of some description is present within the receiver of a communications link, or the read circuit of a storage device. The slicing circuit converts the analogue signals recovered from the various input buffers and equalization circuits of the receiver into discrete levels, 0 and 1 in the case of binary systems. There is, however, information contained within the raw analogue signal that can be of use when performing error correction. With the advent of high-speed analogue-to-digital converters, it is now possible to digitize the incoming bitstream rather than simply slicing it. Instead of a 0 or 1, each input bit can be represented by a number, for example, between zero and seven. A high value such as seven represents a good probability that the bit is a 1, while a low value of, say, zero is a high probability that the bit is 0. Central values around the threshold between levels three and four represent low probabilities of the bit being correct. Typically a bit in error will have a low probability and this allows us to make informed decisions even where the capacity of our correction code has been exceeded. The final decision as to the correct state of a bit can be decided much later when all bits are considered together, and within the context of the coding scheme. The representation of binary states by probabilities is called soft decision decoding as opposed to hard decision decoding where the decision is made by a hardware slicing circuit.

We can apply the idea of soft decision decoding to the previous example in the event that a double-bit error is suspected, even though d_{min} is only 4. Suppose that errors occurred in bits 3 and 5. The remainder would be $\alpha^3 + \alpha^5 = 00011$. As expected this remainder is not present in Table 5.6, but could be generated by a double error in any of the bit pairs 0/1, 2/7, 3/5, 4/10, 6/14, 8/11 and 9/13. If we have augmented the coding with soft decision thresholding, then we have in effect a probability associated with each bit, or a measure of how likely it is to be correct. By multiplying the probabilities of each bit of the possible error pairs we can see which pair is most likely to be in error since it will have the lowest combined probability.

Table 5.7 Received codes and probabilities before thresholding to binary

						Received codes									
Index	14	13	12	11	10	9	8	7	6	5	4	3	2	1	0
Code	6	2	1	5	1	7	7	0	2	2	1	4	3	1	6
Prob.	3	2	3	2	3	4	4	4	2	2	3	1	1	3	3

Table 5.8 Combined bit pair probabilities

Bit pair	0/1	2/7	3/5	4/10	6/14	8/11	9/13
Combined probability	3×3 (9)	1×4 (4)	1×2 (2)	3×3 (9)	2×3 (6)	4×2 (8)	4×2 (8)

If we consider the (15, 10) message 100101100100001 and add errors into bits 3 and 5 to give 100101100001001, the remainder is 00011 (try this longhand). Table 5.7 shows the received message in terms of codes ranging from 0 (a high probability of 0) to 7 (a high probability of 1), and relative probabilities. The combined probabilities of the bit pairs which can give rise to the remainder 00011 are shown in Table 5.8. From these results it is clear that the bit pair 3/5 is least likely to be correct out of the possible candidates. It turns out that the use of soft decisions within an error-correcting code can give as much as 2 dB (slightly more in theory) signal-to-noise ratio advantage over a simple hard decision scheme.

5.6 DISCUSSION

You may not have realized it yet but you have now been exposed to finite field algebra both with error detection using the CRC, and error correction here. This is essentially the same maths that is used in the very powerful techniques that we will be studying next, and hopefully you won't have found it as fearsome as you might have expected. The translation of single-bit errors into remainders which form the finite field defined by the generator polynomial, demonstrates some of the properties of these fields as does the translation of multiple-bit errors into remainders representing the summations of remainders from the individual bit errors.

You should be able to locate a single-bit error given a non-zero remainder at the receiver, and you should also be able to adapt your CRC generator/checker to perform this location. Also you will be able to work out the maximum message size for a given CRC bit length for which correction is still possible, and where messages are significantly smaller than this, determine some instances where multiple-bit errors have occurred. In addition to this, you will be able to modify your generator polynomial so as to be able to distinguish between single- and double-bit errors.

Part Three

Time domain Reed–Solomon coding

6

Correcting errors in the time domain

Reed–Solomon codes are a special example of a more general class of block codes called BCH codes after Bose, Chaudhuri and Hocquenghem, forefathers of the theory. Reed–Solomon (RS) error correction can be understood and implemented in a variety of ways, the principal ones being time domain and frequency domain. Time domain coding is easy to grasp but rather limited in application, while frequency domain coding is perhaps a little harder to grasp but its application is much more amenable to generalized solutions. We can of course combine both of these to yield a third, 'best of both worlds' solution in some cases, but more of this later.

6.1 MANIPULATING FINITE FIELD ELEMENTS

RS coding makes extensive use of finite fields, and for simplicity we will mostly use $GF(2^3)$ since this brings worked examples down to manageable proportions. $GF(2^3)$ has already been calculated in Table 3.1 using the primitive polynomial $x^3 + x + 1 = 0$. At this point it might be useful to outline a few of the operations that we can perform on finite field elements. The elements are listed again in Table 6.1.

Table 6.1 GF(2^3)

Power	Binary value	Decimal value
α	010	2
α^2	100	4
α^3	011	3
α^4	110	6
α^5	111	7
α^6	101	5
α^7	001	1

6.1.1 Multiplying and dividing elements

For any finite field, if $n = 2^m - 1$, $\alpha^n = 1 = \alpha^0$. In this example $\alpha^7 = 1$ and this is quite a useful result. Multiplying and dividing elements are perhaps the easiest operations and can be considered in two ways.

1. $\alpha^2\alpha^4 \qquad = \alpha^6$
   $\alpha^5\alpha^6 \qquad = \alpha^{11} = \alpha^7\alpha^4 = (1)\alpha^4 \ = \alpha^4$

2. $\alpha^2\alpha^4 \qquad = (x^2)(x^2 + x) \ = x^4 + x^3 = x^3(x + 1) = (x + 1)(x + 1) = x^2 + 1$
   $\qquad\qquad = \alpha^6$

The second solution serves mainly to demonstrate that we end up with the same result although hardware realizations of this kind of maths may well make use of this property, dividing works in much the same way, as shown below.

$$\alpha^4/\alpha^2 = \alpha^2$$
$$\alpha^2/\alpha^4 = (1)\alpha^2/\alpha^4 = \alpha^7\alpha^2/\alpha^4 = \alpha^5$$

For evidence of the latter, try multiplying α^5 and α^4 longhand to get α^2.

6.1.2 Addition

Addition is actually the same as subtraction and works as follows:

$$\alpha^5 + \alpha^6 \quad = \quad 111 \oplus$$
$$\underline{101}$$
$$= \quad 010 \ = \alpha \quad (\oplus = \text{XOR})$$

$$\alpha^2 + \alpha^7 \quad = \quad 100 \oplus$$
$$\underline{001}$$
$$= \quad 101 \ = \alpha^6$$

With a little practice, manipulating these elements becomes very straightforward. Like Hamming codes these codes are also linear. Linearity and distributivity are easily demonstrated by considering $\alpha(\alpha + \alpha^2)$. The final outcome is the same whether the bracketed terms are added first and then multiplied by α, or multiplied out first and then added.

6.1.3 Roots

It is possible in some cases to calculate square and cube roots, etc., of finite field elements, but beware because not all fields yield a valid result for all powers. The easiest way to calculate these relationships is to work from the answer back to the question. If we want to know say $\sqrt{\alpha^3}$, then we need to find what n gives $\alpha^{2n} = \alpha^3$. Table 6.2 below gives some examples.

Table 6.2 Examples of roots over GF(2^3)

α^n	α^{2n}	$\alpha^{n/2}$	α^{3n}	$\alpha^{n/3}$	α^{4n}	$\alpha^{n/4}$	α^{5n}	$\alpha^{n/5}$	α^{6n}	$\alpha^{n/6}$
α^1	α^2	α^4	α^3	α^5	α^4	α^2	α^5	α^3	α^6	α^6
α^2	α^4	α^1	α^6	α^3	α^1	α^4	α^3	α^6	α^5	α^5
α^3	α^6	α^5	α^2	α^1	α^5	α^6	α^1	α^2	α^4	α^4
α^4	α^1	α^2	α^5	α^6	α^2	α^1	α^6	α^5	α^3	α^3
α^5	α^3	α^6	α^1	α^4	α^6	α^3	α^4	α^1	α^2	α^2
α^6	α^5	α^3	α^4	α^2	α^3	α^5	α^2	α^4	α^1	α^1
α^7	α^7	α^7	α^7	α^7	α^7	α^7	α^7	α^7	α^7	α^7

Over specific field sizes it is possible to generate simple expressions for the calculation of the roots. For example the $\sqrt{\alpha^n}$ is simply α^{4n}. You will actually find that all of the root columns have an associated column of powers which is identical, making the solution of all roots simple in this field. Over GF(2^3), there are solutions to all roots, but if we take a larger field like GF(2^8), it is easy to demonstrate that this result is not a general one. Try this for yourself, considering α^5 and $\sqrt[5]{\alpha}$ over GF(2^8).

We won't actually be needing these particular results for many of the coding examples considered but they are nonetheless useful identities.

6.1.4 Paired primitive polynomials

You will have seen from previous examples that the hardware circuits used to generate a finite field given a primitive polynomial, actually multiply their register contents by α for each clock pulse so creating the finite field sequence. When considering location of a single-bit error using a non-zero CRC remainder, it was suggested that counting down by multiplying the remainder by α^{-1} would be more appropriate than counting up then subtracting. For GF(2^3) α^{-1} is actually α^6 since $\alpha^{-1} = \alpha^{-1}.1 = \alpha^{-1}.\alpha^7 = \alpha^6$. If we take any non-zero member of the field and successively multiply it by α^{-1} then we will produce all non-zero members of the field, but in reverse order. We can actually use this result in a roundabout sort of way (and with a little stretching of the imagination) to create a second primitive polynomial.

If we have a primitive polynomial, say $x^3 + x + 1 = 0$, used in these examples then the bit pattern of α^{-1} can be found by shifting all the bits of the polynomial right and discarding the least significant bit which drops out of the end. So $\alpha^{-1} = (1011)$ SHR $1 = 101$. We can verify this result from Table 6.1. By flipping this bit pattern and adding 1 to the left-hand side, we create a second primitive polynomial, in this case 1101. If we take a larger polynomial, $285 = 1\ 0\ 0\ 0\ 1\ 1\ 1\ 0\ 1$, then $\alpha^{-1} = 1\ 0\ 0\ 0\ 1\ 1\ 1\ 0$ so the paired polynomial is $1\ 0\ 1\ 1\ 1\ 0\ 0\ 0\ 1 = 369$. If you consult Appendix A you will find this listed among the primitive polynomials of GF(2^8). The result of this operation is, in fact, simply a bit reversal of the complete primitive polynomial. While paired by virtue of this property, the two sequences are not similar in appearance (i.e. one being the reverse). This result is included mostly for the experimenter, but it does allow a speed up in the generation of polynomials over large fields.

6.1.5 Zech logarithms

Addition and multiplication of elements takes place in different domains. With addition we must deal with the bit patterns of the elements, while multiplication which can be done on the basis of pattern is more readily performed using the power index of the element. Because these domains are as different as apples and bananas it is hard to form analytical solutions to some problems. Zech logarithms assist addition as follows:

$$\alpha^{z(n)} = \alpha^n + 1$$

To perform $\alpha^a + \alpha^b$, the expression is rearranged as $\alpha^a(\alpha^{b-a} + 1)$ which equals $\alpha^a \alpha^{z(b-a)}$. This manipulation still requires some form of look-up table but is likely to be slightly faster than translating powers to patterns, adding and translating back to powers, though. See if you can work out the Zech logarithms for GF(2^3).

6.1.6 Solution to a quadratic

Later as we consider locating errors it will become apparent that the solutions to problems often lie in being able to solve a quadratic. For example, the expression

$$\left(\alpha^i + \alpha^a\right)\left(\alpha^i + \alpha^b\right) = 0$$

has two solutions

$$\alpha^i = \alpha^a \text{ and } \alpha^i = \alpha^b$$

The general form of a quadratic equation will thus be

$$\alpha^{2i} + \alpha^{i+P} + \alpha^q = 0$$

Expanding the quadratic we know that $\alpha^P = \alpha^a + \alpha^b$ and $\alpha^q = \alpha^{a+b}$ but this involves solving two simultaneous equations that are in very different domains. $\alpha^P = \alpha^a + \alpha^b$ gives us a problem based in bit patterns while $\alpha^q = \alpha^{a+b}$ gives us a problem whose solution lies in powers. This poses a somewhat more complicated problem than might first be expected. If we apply classic substitution techniques, then we end up with the same problem that we started with. Consider the following.

$$\alpha^b = \alpha^P + \alpha^a$$

so by substitution

$$\alpha^q = \alpha^a\left(\alpha^a + \alpha^P\right)$$

Rearranging gives

$$\alpha^{2a} + \alpha^{a+P} + \alpha^q = 0$$

Lo and behold, we are solving the same problem again. As they say, 'for the definition of recursion, see recursion'. To actually solve this problem we can tackle it in two ways. The first is by trial and error, which is not as bad as it sounds. For an example, let $p = 1$ and $q = 3$. This gives us the expression

$$\alpha^{2i} + \alpha^{i+1} + \alpha^3 = 0$$

Table 6.3 Evaluating the two solutions of a quadratic

Index	*Equation*	*Result*	*= 0 ?*
$i = 0$	$\alpha^7 + \alpha^1 + \alpha^3$	0	✓
$i = 1$	$\alpha^2 + \alpha^2 + \alpha^3$	α^3	✗
$i = 2$	$\alpha^4 + \alpha^3 + \alpha^3$	α^4	✗
$i = 3$	$\alpha^6 + \alpha^4 + \alpha^3$	0	✓
$i = 4$	$\alpha^1 + \alpha^5 + \alpha^3$	α^6	✗
$i = 5$	$\alpha^3 + \alpha^6 + \alpha^3$	α^6	✗
$i = 6$	$\alpha^5 + \alpha^7 + \alpha^3$	α^6	✗
$i = 7$	$\alpha^7 + \alpha^1 + \alpha^3$	0	✓

Now $i = 0$ is the same result as $i = 7$ so we have found the two roots. This may get a little slow over large fields, but with a bit of care, simplification of the problem is possible. Since $\alpha^p = \alpha^a + \alpha^b$, and $p = 1$ the sum of the two roots must add up to α. The symbol pairs which give this result are 2/4, 3/7 and 5/6. Considering similarly $\alpha^q = \alpha^{a+b}$, the product of the powers must equal α^3. The pairs which give this result are 1/2, 3/7 and 4/6. The only pair common to both domains is 3/7, the answer. This allows us to reduce the search by two-thirds. Suppose we check α^4. From powers 4 is paired with 6 and from patterns, 2. Since the pairs don't match, there is now no point in checking α^2 or α^6.

The primitive polynomial provides a link between patterns and powers which allows consideration of a more analytical approach. If we describe the solutions in the form

$$\alpha^a = cx^2 + dx + ex^0$$

and

$$\alpha^b = fx^2 + gx + hx^0$$

where c, d, e, f, g and h are either 0 or 1, then

$$\alpha^p = x^2(c \oplus f) + x(d \oplus g) + x^0(e \oplus h)$$

and

$$\alpha^q = x^4(c.f) + x^3(c.g \oplus d.f) + x^2(c.h \oplus d.g \oplus e.f) + x(d.h \oplus e.g) + x^0(e.h)$$

Clearly this field can't contain bits in powers of greater than x^2, so using the polynomial, $x^4 = x^2 + x$ and $x^3 = x + 1$. This allows us to redistribute these higher terms amongst the lower ones to give

$$\alpha^q = x^2(c.h \oplus d.g \oplus e.f \oplus c.f)$$
$$+ x(d.h \oplus e.g \oplus c.f \oplus c.g \oplus d.f)$$
$$+ x^0(e.h \oplus c.g \oplus d.f)$$

Since we are dealing with Boolean algebra in some of these equations rather than symbol manipulation, the $\oplus$ symbol has been used for bitwise XORing in order to distinguish its function from the OR operation which is usually denoted by +.

Using $\alpha^p = 010$ ($p = 1$) and $\alpha^q = 011$ ($q = 3$), we can form six equations as follows:

from p

$c \oplus f = 0$	so	$f = c$	(from x^2 bits)
$d \oplus g = 1$	so	$g = \bar{d}$	(from x^1 bits)
$e \oplus h = 0$	so	$h = e$	(from x^0 bits)

from q

$c.h \oplus d.g \oplus e.f \oplus c.f$	$= 0$	(from x^2 bits)
$d.h \oplus e.g \oplus c.f \oplus c.g \oplus d.f$	$= 1$	(from x^1 bits)
$e.h \oplus c.g \oplus d.f$	$= 1$	(from x^0 bits)

Substituting to eliminate f, g and h leaves

$c.e \oplus d.(\bar{d}) \oplus (e \oplus c).c = 0$		(from x^2 bits)
$d.e \oplus (e \oplus c).(\bar{d}) \oplus (c \oplus d).c = 1$		(from x^1 bits)
$e.e \oplus c.(\bar{d}) \oplus d.c = 1$		(from x^0 bits)

Now $n^2 = n$ since $1.1 = 1$ and $0.0 = 0$ and $n.\bar{n} = 0$ since $0.1 = 0$ and $1.0 = 0$ so considering x^2 and expanding gives

$$c.e \oplus 0 \oplus c.c \oplus e.c = 0$$

The common $e.c$ ($=c.e$) term cancels leaving $c = 0$. Expanding x terms (and substituting $1 \oplus d$ for $\bar{d}$) gives

$$d.e \oplus e \oplus c \oplus d.e \oplus d.c \oplus c.c \oplus d.c = 1$$

Cancelling the common $d.e$ and $d.c$ and c terms leaves $e = 1$. From x^0 terms (again substituting $1 \oplus d$ for $\bar{d}$)

$$e \oplus c \oplus c.d \oplus d.c = 1$$

Cancelling common terms leaves $e \oplus c = 1$ or $c = \bar{e}$ so

$$c = 0 \text{ (from } x^2 \text{ bits)},$$
$$e = 1 \text{ (from } x^1 \text{ bits) and}$$
$$c = \bar{e} \text{ (from } x^0 \text{ bits)}$$

This tells us that the solution to $\alpha^a = 0?1$ and we appear to have no information about d, the middle bit. In fact, our distinction between the two solutions, α^a and α^b, is quite notional. One would initially think that we must first solve for c, d and e, then use these to solve f, g and h. However, because this distinction is notional, c, d and e actually reveal both solutions as would f, g and h if we had chosen to solve them instead. We know that the answers are 011 and 001. These both share common most significant and least significant bits and therefore we arrived earlier at definitive solutions for them, i.e. $c = 0$ and $e = 1$. Because d shares both states (0 and 1) in the two solutions, we arrive at no fixed result. Our two solutions must therefore be 001 and 011.

To reinforce this idea, using $\alpha^a = 101$ and $\alpha^b = 010$ we'll repeat the calculations. In other words all bits are different. α^p is, therefore, 111 while α^q is 001. From p

$c \oplus f = 1$	or	$f = \bar{c}$	(from x^2 bits)
$d \oplus g = 1$	or	$g = \bar{d}$	(from x^1 bits)
$e \oplus h = 1$	or	$h = \bar{e}$	(from x^0 bits)

from q

$$
\begin{array}{lll}
c.h \oplus d.g \oplus e.f \oplus c.f & = 0 & \text{(from } x^2 \text{ bits)} \\
d.h \oplus e.g \oplus c.f \oplus c.g \oplus d.f & = 0 & \text{(from } x^1 \text{ bits)} \\
e.h \oplus c.g \oplus d.f & = 1 & \text{(from } x^0 \text{ bits)}
\end{array}
$$

Substituting to eliminate f, g and h leaves

$$
\begin{array}{ll}
c.(\bar{e}) \oplus d.(\bar{d}) \oplus (e \oplus c).(\bar{c}) = 0 & \text{(from } x^2 \text{ bits)} \\
d.(\bar{e}) \oplus (e \oplus c).(\bar{d}) \oplus (c \oplus d).(\bar{c}) = 0 & \text{(from } x^1 \text{ bits)} \\
e.(\bar{e}) \oplus c.(\bar{d}) \oplus d.(\bar{c}) = 1 & \text{(from } x^0 \text{ bits)}
\end{array}
$$

so $e = c$ (from x^2 bits), $e = c$ (from x^1 bits) and $c = \bar{d}$ (from x^0 bits).

Hint: To simplify these expressions remember that multiplies (the AND function) take place precedence over adds (the eXclusive-OR function) and where two similar terms appear added, they cancel out. Substituting $(1 \oplus n)$ for $\bar{n}$ can make the reduction easier to follow if you're not overly familiar with Boolean algebra. Take, for example, the expression

$$
e.(\bar{e}) \oplus c.(\bar{d}) \oplus d.(\bar{c}) = 1
$$

Making the substitution, this becomes

$$
e.(1 \oplus e) \oplus c.(1 \oplus d) \oplus d.(1 \oplus c) = 1
$$

so

$$
e \oplus e^2 \oplus c \oplus c.d \oplus d \oplus d.c = 1
$$

and cancelling the common terms leaves

$$
c \oplus d = 1
$$

From these expressions we know that $e = c$ and $d \neq c$. Starting with $c = 0$, then $e = 0$ and $d = 1$ so we have 010. When $c = 1$, then $e = 1$ and $d = 0$ giving 101, so we find our two roots again. In general where we reach a definitive result for a bit, it will be true for both roots and where there is no definitive value we will have a relationship which specifies each bit with respect to the other bits.

For a given polynomial, we can specify a set of equations which will lead to the roots. Let our known information

$$\alpha^p = p_2 x^2 + p_1 x + p_0 x^0$$
$$\alpha^q = q_2 x^2 + q_1 x + q_0 x^0$$

and our required roots be two solutions for

$$\alpha^a = a_2 x^2 + a_1 x + a_0 x^0$$
$$\alpha^b = b_2 x^2 + b_1 x + b_0 x^0$$

where p_n, q_n, a_n and b_n are binary bits (1 or 0). Using the equations derived earlier for the polynomial $x^3 + x + 1 = 0$ we have

$$a_2(a_0 \oplus p_0) \oplus a_1(a_1 \oplus p_1) \oplus (a_2 \oplus a_0)(a_2 \oplus p_2) = q_2$$
$$a_1(a_0 \oplus p_0) \oplus (a_2 \oplus a_0)(a_1 \oplus p_1) \oplus (a_2 \oplus a_1)(a_2 \oplus p_2) = q_1$$
$$a_0(a_0 \oplus p_0) \oplus a_2(a_1 \oplus p_1) \oplus a_1(a_2 \oplus p_2) = q_0$$

Using the previous example, $\alpha^p = 111$ so $p_2 = 1$, $p_1 = 1$ and $p_0 = 1$ and $\alpha^q = 001$, so $q_2 = 0$, $q_1 = 0$ and $q_0 = 1$. Substituting these into the above gives

$$a_2(a_0 \oplus 1) \oplus a_1(a_1 \oplus 1) \oplus (a_2 \oplus a_0)(a_2 \oplus 1) = 0$$
$$a_1(a_0 \oplus 1) \oplus (a_2 \oplus a_0)(a_1 \oplus 1) \oplus (a_2 \oplus a_1)(a_2 \oplus 1) = 0$$
$$a_0(a_0 \oplus 1) \oplus a_2(a_1 \oplus 1) \oplus a_1(a_2 \oplus 1) = 1$$

which simplifies to

$$a_2 \oplus a_0 = 0 \text{ or } a_2 = a_0$$
$$a_2 \oplus a_0 = 0 \text{ or } a_2 = a_0$$
$$a_2 \oplus a_1 = 1 \text{ or } a_2 = \bar{a}_1$$

This, of course, is the same result that we arrived at before, giving the solutions 010 and 101. Where there are two equal roots, the α^p term is 0 so we have the identity

$$\left(\alpha^i + \alpha^a\right)^2 = \alpha^{2i} + \alpha^{2a}$$

Using the previous notation, then if $\alpha^p = 0$, $\alpha^a = \sqrt{\alpha^q}$ which can be determined from Table 6.2.

6.2 INTRODUCTION TO THE TIME DOMAIN

The use of the term 'time domain' comes from an analogy with signal processing. In general signals are expressed and seen in the time domain, for example, a sine wave on an oscilloscope. Another representation of the signal, however, is in the frequency domain where a sine wave would appear as a single point in two-dimensional space describing frequency and amplitude. Signals usually exist in the time domain and have to undergo transformation in order to be seen in the frequency domain. It is at this point that the analogy ends. With time domain error coding the encoded message contains the protected data in its original format. If we were to eavesdrop onto a digital network, time domain ASCII messages would be quite visible in their original format plus error codes and framework bits. In other words, the data is transmitted as is, with additional protection coding appended.

In the frequency domain (discussed later) the data undergoes transformation before being launched onto the communication or storage medium. This is not usually for the sake of encryption, but convenience of coding. In this case, the eavesdropper would see nothing recognizable without first converting the message back into the time domain.

6.3 CALCULATING CHECK SYMBOLS FOR ONE ERROR

With this kind of error correction we deal with symbol errors rather than just bit errors. A single-symbol error-correcting code will be able to correct any number of bit errors within a single symbol. The symbols have one fewer bits than the generator polynomial. For $GF(2^m)$ this is m bits. Given this is the case, when a single error occurs we have two unknowns, the symbol position of the error within the message, and the bit pattern of the error. As with any problem, to solve two unknowns we must develop two simultaneous equations, and this can be done as follows. For $GF(2^m)$ the data are split into blocks of $2^m - 1$ symbols, each symbol m bits long. Over $GF(2^3)$ the block will comprise seven three-bit symbols. We will call the first five symbols A, B, C, D and E, which represent 15 bits of data. The last two symbols we'll call R and S. These are our Reed–Solomon symbols which are as yet unknown.

We can generate the two simultaneous equations in a variety of ways and the following is just one example:

$$A + B + C + D + E + R + S = 0 \qquad (6.1)$$

$$\alpha^1 A + \alpha^2 B + \alpha^3 C + \alpha^4 D + \alpha^5 E + \alpha^6 R + \alpha^7 S = 0 \qquad (6.2)$$

Initially, these will seem a bit arbitrary, but if we set R and S such that (6.1) and (6.2) are satisfied, we have placed two roots in the message vector. If at the receiver the message does not contain two roots, we know that an error has occurred. Careful construction of (6.1) and (6.2) means that if only a single-symbol error occurs, then we can correct it. First we need to solve them to find R and S, based on A to E.

Rearranging (6.1) and substituting into (6.2) we get

$$S = A + B + C + D + E + R$$

so

$$\alpha^1 A + \alpha^2 B + \alpha^3 C + \alpha^4 D + \alpha^5 E + \alpha^6 R + \alpha^7 (A + B + C + D + E + R) = 0$$

Gathering up the terms (and remembering that $\alpha^7 = 1$),

$$R(\alpha^6 + 1) = A(\alpha^1 + 1) + B(\alpha^2 + 1) + C(\alpha^3 + 1) + D(\alpha^4 + 1) + E(\alpha^5 + 1)$$

Solving the additions (e.g. $\alpha^2 + 1 = 100 \oplus 001 = 101$),

$$R(\alpha^2) = A(\alpha^3) + B(\alpha^6) + C(\alpha^1) + D(\alpha^5) + E(\alpha^4)$$

and dividing out

$$\boldsymbol{R = A\alpha + B\alpha^4 + C\alpha^6 + D\alpha^3 + E\alpha^2}$$

Substituting back for S,

$$S = A + B + C + D + E + A\alpha + B\alpha^4 + C\alpha^6 + D\alpha^3 + E\alpha^2$$

$$S = A(\alpha^1 + 1) + B(\alpha^4 + 1) + C(\alpha^6 + 1) + D(\alpha^3 + 1) + E(\alpha^2 + 1)$$

$$\boldsymbol{S = A\alpha^3 + B\alpha^5 + C\alpha^2 + D\alpha + E\alpha^6}$$

Having found R and S in terms of A, B, C, D and E, we now need to see how this result can be used to find and locate errors. At the receiver two syndromes S_0 and S_1 are calculated using (6.3) and (6.4):

$$S_0 = A + B + C + D + E + R + S \qquad (6.3)$$

$$S_1 = \alpha^1 A + \alpha^2 B + \alpha^3 C + \alpha^4 D + \alpha^5 E + \alpha^6 R + \alpha^7 S \qquad (6.4)$$

If no errors have occurred then both S_0 and S_1 will be zero since we arranged for this to happen when we calculated R and S. Suppose, however, that symbol B gets corrupted by an error with a pattern ε, such that the receiver sees $B' = B + \varepsilon$. The syndromes will now take on the following values:

$$S_0 = A + (B + \varepsilon) + C + D + E + R + S$$
$$= (A + B + C + D + E + R + S) + \varepsilon$$

$$S_1 = \alpha^1 A + \alpha^2 (B + \varepsilon) + \alpha^3 C + \alpha^4 D + \alpha^5 E + \alpha^6 R + \alpha^7 S$$
$$= (\alpha^1 A + \alpha^2 B + \alpha^3 C + \alpha^4 D + \alpha^5 E + \alpha^6 R + \alpha^7 S) + \alpha^2 \varepsilon$$

The bracketed parts of the above are zero since we arranged this when we calculated R and S. This leaves S_0 with the value ε, and S_1 with $\alpha^2 \varepsilon$. In other words, S_0 holds the error pattern while S_1 holds the error pattern times a locating value. To locate the error we simply evaluate k where

$$k = S_1/S_0 = \alpha^2 \varepsilon / \varepsilon = \alpha^2$$

k tells us where the error is, i.e. we know that data symbol B is multiplied by α^2 so this is the location of the error. To correct the error we take the received symbol B' and add to it S_0 such that

$$B = B' + \varepsilon$$

6.4 EXAMPLE OF A ONE-SYMBOL ERROR CORRECTION

Let the data be as follows:

$$A = 110\ (\alpha^4),\ B = 000\ (0),\ C = 010\ (\alpha),\ D = 100\ (\alpha^2),\ E = 111\ (\alpha^5)$$

Now

$$R = A\alpha + B\alpha^4 + C\alpha^6 + D\alpha^3 + E\alpha^2$$

and

$$S = A\alpha^3 + B\alpha^5 + C\alpha^2 + D\alpha + E\alpha^6$$

so

$$R = \alpha^4\alpha + 0\alpha^4 + \alpha\alpha^6 + \alpha^2\alpha^3 + \alpha^5\alpha^2 = \alpha^5 + 0 + \alpha^7 + \alpha^5 + \alpha^7 = 0$$

$$S = \alpha^4\alpha^3 + 0\alpha^5 + \alpha\alpha^2 + \alpha^2\alpha + \alpha^5\alpha^6 = \alpha^7 + 0 + \alpha^3 + \alpha^3 + \alpha^4 = \alpha^5$$

The transmitted data sequence is 110, 000, 010, 100, 111, 000, 111.

To check for errors we calculate the two syndromes S_0 and S_1 at the receiver. This is done by adding the '1's in the columns modulo-2 in Table 6.4.

Table 6.4 Checking the syndromes (no errors)

Symbol	Bits			Symbol	Bits		
A	1	1	0	$\alpha^1.A$	1	1	1
B	0	0	0	$\alpha^2.B$	0	0	0
C	0	1	0	$\alpha^3.C$	1	1	0
D	1	0	0	$\alpha^4.D$	1	0	1
E	1	1	1	$\alpha^5.E$	0	1	1
R	0	0	0	$\alpha^6.R$	0	0	0
S	1	1	1	$\alpha^7.S$	1	1	1
S_0	0	0	0	S_1	0	0	0

Table 6.5 Finding the syndromes after an error

Symbol	Bits			Symbol	Bits		
A	1	1	0	$\alpha^1.A$	1	1	1
B	0	0	0	$\alpha^2.B$	0	0	0
C	0	1	0	$\alpha^3.C$	1	1	0
D	1	0	0	$\alpha^4.D$	1	0	1
E'	1	0	0	$\alpha^5.E'$	0	0	1
R	0	0	0	$\alpha^6.R$	0	0	0
S	1	1	1	$\alpha^7.S$	1	1	1
S_0	0	1	1	S_1	0	1	0

In the absence of errors the syndromes are zero just as we arranged when we set R and S. Suppose now that the error $\varepsilon = 011$ (α^3) corrupts symbol E such that $E' = 100$ (α^2). The syndromes are re-evaluated in Table 6.5. Since the syndromes are not zero, we know that an error has occurred. $S_0 = \alpha^3$ and $S_1 = \alpha$ so k, the locator, is (α^7)$\alpha/\alpha^3 = \alpha^5$ and the error $\varepsilon = S_0 = \alpha^3$. The locator k points us to symbol E', so the corrected symbol is simply

$$E = E' \oplus S_0 = 100 \oplus 011 = 111$$

6.5 DISCUSSION

At this point you should be able to construct the starting conditions for a single-correcting Reed–Solomon encoded block of data. You have seen that to correct one (symbol) error, there are now two unknowns, position and pattern, requiring two check symbols for their calculation.

It should be stressed that the two starting conditions given in (6.1) and (6.2) are by no means the only ones. Provided we do not use the GF elements twice, the ordering of the symbols is arbitrary and can be selected for convenience of decoding. We do, however, need to be a little careful if we wish to extend these ideas to correcting more than a single symbol error.

You will also have a good idea of some of the ways in which finite field elements can be manipulated, including adding, multiplying, raising to powers and solving quadratics.

7

Correcting two symbols

7.1 CORRECTING BY ERASURE

With two-symbol errors there are four unknowns, two error patterns and two error positions. Four unknowns require the solution of four simultaneous equations and two equations are required to correct one error. Before we look at how this can be arranged it is worth examining a process of correction called **erasure**. As we will see later under data packaging, there are occasions when the positions of the errors may be available even though their pattern is not. If this is the case then the two simultaneous equations used before to find one error (pattern and position) can be used to find two error patterns or, more generally, any **two** unknowns.

Using the example of the previous chapter, suppose it is known that data symbol A and check symbol S have been corrupted. To correct the message the received symbols A and S are erased by replacing them with zeros, and the syndromes are calculated as in Table 7.1.

Table 7.1 Correcting two symbols by erasure

Symbol	Bits			Symbol	Bits		
A	0	0	0	$\alpha^1 A$	0	0	0
B	0	0	0	$\alpha^2 B$	0	0	0
C	0	1	0	$\alpha^3 C$	1	1	0
D	1	0	0	$\alpha^4 D$	1	0	1
E	1	1	1	$\alpha^5 E$	0	1	1
R	0	0	0	$\alpha^6 R$	0	0	0
S	0	0	0	$\alpha^7 S$	0	0	0
S_0	0	0	1	S_1	0	0	0

Since

$$0 = A + B + C + D + E + R + S$$

and

$$S_0 = 0 + B + C + D + E + R + 0$$

then

$$S_0 + A + S = A + B + C + D + E + R + S = 0$$

leaving

$$S_0 = A + S$$

Similarly

$$S_1 = \alpha A + \alpha^7 S$$

Rearranging and substituting (remembering also that $\alpha^7 = 1$) gives

$$S_1 = \alpha A + S_0 + A$$

so

$$A = (S_1 + S_0)/(\alpha + 1)$$
$$= (S_1 + S_0)/\alpha^3$$
$$= (S_1 + S_0)\alpha^4$$

Putting in values of S_0 and S_1 from Table 7.1 gives

$$A = (000 + 001).\alpha^4 = \alpha^4 = 110$$

Substituting back gives

$$S = S_0 + A = 001 + 110 = 111$$

So we have recovered the original symbols A and S.

7.2 FINDING CHECK SYMBOLS FOR TWO ERRORS

For four unknowns, four orthogonal equations are constructed as follows:

$$A + B + C + P + Q + R + S = 0 \tag{7.1}$$

$$A\alpha^7 + B\alpha^6 + C\alpha^5 + P\alpha^4 + Q\alpha^3 + R\alpha^2 + S\alpha^1 = 0 \tag{7.2}$$

$$A\alpha^7 + B\alpha^5 + C\alpha^3 + P\alpha^1 + Q\alpha^6 + R\alpha^4 + S\alpha^2 = 0 \tag{7.3}$$

$$A\alpha^7 + B\alpha^4 + C\alpha^1 + P\alpha^5 + Q\alpha^2 + R\alpha^6 + S\alpha^3 = 0 \tag{7.4}$$

where A, B and C are data, and P, Q, R and S are check symbols.

The term orthogonal loosely applies to these equations in the sense that each symbol is associated with a unique combination of GF elements. To establish these four equations we clearly need four check symbols, leaving only three data symbols. While this is of little value over seven symbol messages, it becomes increasingly economic for longer messages. To illustrate the relative unimportance of the ordering of GF symbols, they are multiplied by the message in ascending powers (towards the left), rather than descending in the previous example. In (7.1) all symbols are multiplied by α^0, increasing in power by 0. In (7.2), the symbols start at α and increase by powers of 1. In (7.3), the powers increase by 2 and so on. If we term the symbol positions i ($1 \le i \le 7$ leftwards) then for (7.1), each symbol is multiplied by α^{0i}, for (7.2) α^i, for (7.3) α^{2i}, and for (7.4) α^{3i}.

Rearranging (7.1) gives

$$P = A + B + C + Q + R + S$$

hence

$$\alpha^4 P = \alpha^4(A + B + C + Q + R + S)$$

Combining (7.1) and (7.2) gives

$$A(\alpha^7 + \alpha^4) + B(\alpha^6 + \alpha^4) + C(\alpha^5 + \alpha^4) + Q(\alpha^3 + \alpha^4)$$
$$+ R(\alpha^2 + \alpha^4) + S(\alpha + \alpha^4) = 0$$

$$A(001 + 110) + B(101 + 110) + C(111 + 110) + Q(011 + 110)$$
$$+ R(100 + 110) + S(010 + 110) = 0$$

$$A(111) + B(011) + C(001) + Q(101) + R(010) + S(100) = 0$$

$$A\alpha^5 + B\alpha^3 + C\alpha^7 + Q\alpha^6 + R\alpha + S\alpha^2 = 0 \tag{7.5}$$

Combining (7.1) and (7.3) gives

$$A(\alpha^7 + \alpha) + B(\alpha^5 + \alpha) + C(\alpha^3 + \alpha) + Q(\alpha^6 + \alpha)$$
$$+ R(\alpha^4 + \alpha) + S(\alpha^2 + \alpha) = 0$$

$$A\alpha^3 + B\alpha^6 + C\alpha^7 + Q\alpha^5 + R\alpha^2 + S\alpha^4 = 0 \tag{7.6}$$

Combining (7.1) and (7.4) gives

$$A(\alpha^7 + \alpha^5) + B(\alpha^4 + \alpha^5) + C(\alpha + \alpha^5) + Q(\alpha^2 + \alpha^5)$$
$$+ R(\alpha^6 + \alpha^5) + S(\alpha^3 + \alpha^5) = 0$$

$$\boldsymbol{A\alpha^4 + B\alpha^7 + C\alpha^6 + Q\alpha^3 + R\alpha + S\alpha^2 = 0} \qquad (7.7)$$

Rearranging (7.5) gives

$$Q = (A\alpha^5 + B\alpha^3 + C\alpha^7 + R\alpha + S\alpha^2)/\alpha^6$$

$$Q = A\alpha^6 + B\alpha^4 + C\alpha + R\alpha^2 + S\alpha^3$$

Combining (7.5) and (7.6) gives

$$A(\alpha^3 + \alpha^4) + B(\alpha^6 + \alpha^2) + C(\alpha^7 + \alpha^6) + R(\alpha^2 + \alpha^7) + S(\alpha^4 + \alpha) = 0$$

$$\boldsymbol{A\alpha^6 + B\alpha^7 + C\alpha^2 + R\alpha^6 + S\alpha^2 = 0} \qquad (7.8)$$

Combining (7.5) and (7.7) gives

$$A(\alpha^4 + \alpha^2) + B(\alpha^7 + \alpha^7) + C(\alpha^6 + \alpha^4) + R(\alpha + \alpha^5) + S(\alpha^2 + \alpha^6) = 0$$

$$\boldsymbol{A\alpha + C\alpha^3 + R\alpha^6 + S\alpha^7 = 0} \qquad (7.9)$$

Rearranging (7.8) gives

$$R = (A\alpha^6 + B\alpha^7 + C\alpha^2 + S\alpha^2)/\alpha^6$$

$$\boldsymbol{R = A + B\alpha + C\alpha^3 + S\alpha^3} \qquad (7.10)$$

Combining (7.10) and (7.9) gives

$$A\alpha + C\alpha^3 + (A + B\alpha + C\alpha^3 + S\alpha^3)\alpha^6 + S\alpha^7 = 0$$

$$S = (A\alpha^5 + B\alpha^7 + C\alpha^5) / \alpha^6$$

$$\boldsymbol{S = A\alpha^6 + B\alpha + C\alpha^6} \qquad (7.11)$$

Combining (7.11) and (7.10) gives

$$R = A\alpha^6 + B\alpha^2 + C\alpha^5 \qquad (7.12)$$

From (7.5), (7.11) and (7.12)

$$Q = A\alpha^3 + B\alpha^4 + C\alpha^5 \qquad (7.13)$$

From (7.1), (7.11), (7.12) and (7.13)

$$P = A\alpha + B\alpha^7 + C\alpha^2 \qquad (7.14)$$

Equations (7.11) to (7.14) thus describe how to generate P, Q, R and S in order to satisfy the four starting conditions set by (7.1) to (7.4), based on three data symbols A, B and C. To demonstrate the generation of the check symbols, consider the data

$$A = 011 \ (\alpha^3), \ B = 101 \ (\alpha^6), \ C = 111 \ (\alpha^5)$$

$$
\begin{aligned}
P &= \alpha^4 + \alpha^6 + \alpha^7 = \alpha && (110 \oplus 101 \oplus 001 = 010) \\
Q &= \alpha^6 + \alpha^3 + \alpha^3 = \alpha^6 && (101 \oplus 011 \oplus 011 = 101) \\
R &= \alpha^2 + \alpha \ + \alpha^3 = \alpha^6 && (100 \oplus 010 \oplus 011 = 101) \\
S &= \alpha^2 + \alpha^7 + \alpha^4 = \alpha^3 && (100 \oplus 001 \oplus 110 = 011)
\end{aligned}
$$

The syndromes S_0 to S_3 are calculated in Table 7.2 below.

Table 7.2 Checking the syndromes (no errors)

Evaluating S_0			*Evaluating S_1*			*Evaluating S_2*			*Evaluating S_3*		
A	α^3	011	$A\alpha^7$	α^3	011	$A\alpha^7$	α^3	011	$A\alpha^7$	α^3	011
B	α^6	101	$B\alpha^6$	α^5	111	$B\alpha^5$	α^4	110	$B\alpha^4$	α^3	011
C	α^5	111	$C\alpha^5$	α^3	011	$C\alpha^3$	α^1	010	$C\alpha^1$	α^6	101
P	α^1	010	$P\alpha^4$	α^5	111	$P\alpha^1$	α^2	100	$P\alpha^5$	α^6	101
Q	α^6	101	$Q\alpha^3$	α^2	100	$Q\alpha^6$	α^5	111	$Q\alpha^2$	α^1	010
R	α^6	101	$R\alpha^2$	α^1	010	$R\alpha^4$	α^3	011	$R\alpha^6$	α^5	111
S	α^3	011	$S\alpha^1$	α^4	110	$S\alpha^2$	α^5	111	$S\alpha^3$	α^6	101
S_0		000	S_1		000	S_2		000	S_3		000

So all the syndromes check out, giving the expected 0 result. Now let's introduce two errors, ε_i and ε_j, into the data at positions i and j. The syndromes will now have the values

$$S_0 = \varepsilon_i + \varepsilon_j \tag{7.15}$$

$$S_1 = \varepsilon_i \alpha^i + \varepsilon_j \alpha^j \tag{7.16}$$

$$S_2 = \varepsilon_i \alpha^{2i} + \varepsilon_j \alpha^{2j} \tag{7.17}$$

$$S_3 = \varepsilon_i \alpha^{3i} + \varepsilon_j \alpha^{3j} \tag{7.18}$$

Rearranging (7.15) gives

$$\varepsilon_i = S_0 + \varepsilon_j$$

Substituting into (7.16), (7.17) and (7.18) gives

$$S_1 = (S_0 + \varepsilon_j)\alpha^i + \varepsilon_j \alpha^j \tag{7.19}$$

$$S_2 = (S_0 + \varepsilon_j)\alpha^{2i} + \varepsilon_j \alpha^{2j} \tag{7.20}$$

$$S_3 = (S_0 + \varepsilon_j)\alpha^{3i} + \varepsilon_j \alpha^{3j} \tag{7.21}$$

Rearranging (7.19) gives

$$\varepsilon_j(\alpha^i + \alpha^j) = S_1 + S_0\alpha^i$$

$$\varepsilon_j = (S_1 + S_0\alpha^i)/(\alpha^i + \alpha^j)$$

Substituting into (7.20) and (7.21) gives

$$S_2 = (S_1 + S_0\alpha^i)/(\alpha^i + \alpha^j).(\alpha^{2i} + \alpha^{2j}) + S_0\alpha^{2i}$$

Now

$$(\alpha^{2i} + \alpha^{2j}) = (\alpha^i + \alpha^j)^2$$

so

$$S_2 = (S_1 + S_0\alpha^i)(\alpha^i + \alpha^j) + S_0\alpha^{2i}$$

$$S_2 = S_0\alpha^{i+j} + S_1(\alpha^i + \alpha^j) \tag{7.22}$$

and

$$S_3 = (S_1 + S_0\alpha^i)/(\alpha^i + \alpha^j)(\alpha^{3i} + \alpha^{3j}) + S_0\alpha^{3i}$$

Now

$$(\alpha^{3i} + \alpha^{3j}) = (\alpha^i + \alpha^j)^3 + \alpha^{i+j}(\alpha^i + \alpha^j)$$

so

$$S_3 = (S_1 + S_0\alpha^i)((\alpha^i + \alpha^j)^2 + \alpha^{i+j}) + S_0\alpha^{3i}$$

$$S_3 = S_0\alpha^{i+j}(\alpha^i + \alpha^j) + S_1(\alpha^{2i} + \alpha^{2j} + \alpha^{i+j}) \tag{7.23}$$

Rearranging (7.22) gives

$$\alpha^i(S_0\alpha^j + S_1) = S_2 + S_1\alpha^j$$

$$\alpha^i = (S_2 + S_1\alpha^j)/(S_0\alpha^j + S_1)$$

Substituting into (7.23) gives

$$S_3 = S_0\alpha^j(S_2 + S_1\alpha^j)/(S_0\alpha^j + S_1)((S_2 + S_1\alpha^j)/(S_0\alpha^j + S_1) + \alpha^j)$$
$$+ S_1((S_2 + S_1\alpha^j)^2/(S_0\alpha^j + S_1)^2 + \alpha^{2j} + \alpha^j(S_2 + S_1\alpha^j)/(S_0\alpha^j + S_1))$$

Multiplying by $(S_0\alpha^j + S_1)^2$ gives

$$S_3(S_0\alpha^j + S_1)^2 = S_0\alpha^j(S_2 + S_1\alpha^j)((S_2 + S_1\alpha^j) + \alpha^j(S_0\alpha^j + S_1))$$
$$+ S_1((S_2 + S_1\alpha^j)^2 + \alpha^{2j}(S_0\alpha^j + S_1)^2$$
$$+ \alpha^j(S_2 + S_1\alpha^j)(S_0\alpha^j + S_1))$$

Simplifying gives

$$S_0^2 S_3\alpha^{2j} + S_1^2 S_3 + S_0 S_2^2\alpha^j + S_0 S_1 S_2\alpha^{2j} + S_0 S_1 S_2\alpha^{2j} + S_0 S_1^2\alpha^{3j}$$
$$+ S_0^2 S_2\alpha^{3j} + S_0 S_1 S_2\alpha^{2j} + S_0^2 S_1\alpha^{4j} + S_0 S_1^2\alpha^{3j} + S_1 S_2^2$$
$$+ S_1^3\alpha^{2j} + S_0^2 S_1\alpha^{4j} + S_1^3\alpha^{2j} + S_0 S_1 S_2\alpha^{2j} + S_0 S_1^2\alpha^{3j}$$
$$+ S_1^2 S_2\alpha^j + S_1^3\alpha^{2j} = 0$$

$$\alpha^{3j}(S_0 S_1^2 + S_0^2 S_2) + \alpha^{2j}(S_0^2 S_3 + S_1^3) + \alpha^j(S_0 S_2^2 + S_1^2 S_2)$$
$$+ S_1^2 S_3 + S_1 S_2^2 = 0 \tag{7.24}$$

7.3 EXAMPLE OF A TWO-SYMBOL ERROR CORRECTION

In (7.24), the only unknown is j, the position of ε_j. In actual fact, (7.24) makes no distinction between the notional positions i and j, having solutions representing both i and j. If we introduce two errors, we can see that this solution is correct. Let the first error be at position 7 (A) and the error pattern ε_7 be 100, and the second error be at position 3 (Q) and the error pattern ε_3 be 001. The received symbol A' is thus 111 and Q' is 100. The syndromes are calculated in Table 7.3.

Table 7.3 Calculating the syndromes with two errors

Evaluating S_0			*Evaluating S_1*			*Evaluating S_2*			*Evaluating S_3*		
A	α^5	111	$A\alpha^7$	α^5	111	$A\alpha^7$	α^5	111	$A\alpha^7$	α^5	111
B	α^6	101	$B\alpha^6$	α^5	111	$B\alpha^5$	α^4	110	$B\alpha^4$	α^3	011
C	α^5	111	$C\alpha^5$	α^3	011	$C\alpha^3$	α^1	010	$C\alpha^1$	α^6	101
P	α^1	010	$P\alpha^4$	α^5	111	$P\alpha^1$	α^2	100	$P\alpha^5$	α^6	101
Q	α^2	100	$Q\alpha^3$	α^5	111	$Q\alpha^6$	α^1	010	$Q\alpha^2$	α^4	110
R	α^6	101	$R\alpha^2$	α^1	010	$R\alpha^4$	α^3	011	$R\alpha^6$	α^5	111
S	α^3	011	$S\alpha^1$	α^4	110	$S\alpha^2$	α^5	111	$S\alpha^3$	α^6	101
S_0		101	S_1		111	S_2		001	S_3		000

So

$$S_0 = \alpha^6,\ S_1 = \alpha^5,\ S_2 = \alpha^7 \text{ and } S_3 = 0$$

Substituting these syndromes into (7.24) gives

$$\alpha^{3j}(\alpha^2 + \alpha^5) + \alpha^{2j}\alpha + \alpha^j(\alpha^6 + \alpha^3) + \alpha^5 = 0$$

$$\alpha^{3j}(100 + 111) + \alpha^{2j+1} + \alpha^j(101 + 011) + \alpha^5 = 0$$

$$\alpha^{3j+3} + \alpha^{2j+1} + \alpha^{j+4} + \alpha^5 = 0$$

Dividing through by α^3 gives

$$\alpha^{3j} + \alpha^{2j+5} + \alpha^{j+1} + \alpha^2 = 0$$

The simplest way to solve this is to try all values of j (0 to 7) and see which gives a 0 result.

Table 7.4 Locating the errors

Index	Equation (7.24)	Result	= 0 ?
$j=0$	$\alpha^0 + \alpha^5 + \alpha^1 + \alpha^2$	0	✓
$j=1$	$\alpha^3 + \alpha^7 + \alpha^2 + \alpha^2$	α	✗
$j=2$	$\alpha^6 + \alpha^2 + \alpha^3 + \alpha^2$	α^4	✗
$j=3$	$\alpha^2 + \alpha^4 + \alpha^4 + \alpha^2$	0	✓
$j=4$	$\alpha^5 + \alpha^6 + \alpha^5 + \alpha^2$	α^7	✗
$j=5$	$\alpha^1 + \alpha^1 + \alpha^6 + \alpha^2$	α^7	✗
$j=6$	$\alpha^4 + \alpha^3 + \alpha^7 + \alpha^2$	0	✓
$j=7$	$\alpha^7 + \alpha^5 + \alpha^1 + \alpha^2$	0	✓

From Table 7.4 it appears that there are four solutions or error positions. In fact, $j = 0$ is the same solution as $j = 7$, and therefore it can be ignored. But what of the other three? The two valid errors must also satisfy (7.22) so we can substitute the three possibles into (7.22).

$j = 3$:

$$\alpha^i = (S_2 + S_1\alpha^j)/(S_0\alpha^j + S_1)$$
$$= (\alpha^7 + \alpha^5\alpha^3)/(\alpha^6\alpha^3 + \alpha^5)$$
$$= \alpha^3/\alpha^3 = \alpha^7$$

Since α^7 represents one of the other solutions to (7.24) ($j = 7$) we can take 3 and 7 to be the error positions. If we substitute in $j = 7$ we get back to $i = 3$. Substituting the third solution, $j = 6$, into (7.22) gives the following:

$$\alpha^i = (S_2 + S_1\alpha^j)/(S_0\alpha^j + S_1)$$
$$= (\alpha^7 + \alpha^5\alpha^6)/(\alpha^6\alpha^6 + \alpha^5)$$
$$= \alpha^5/0$$

Clearly this cannot be a valid solution, even in the slightly obscure context of finite field algebra. Since (7.24) must satisfy both solutions, we can replace the α^j term in (7.24) with (7.22). Omitting the maths involved, the solution is

$$\alpha^{2j}(S_0 S_1^4 S_2 + S_0^2 S_1^2 S_2^2 + S_1^6 + S_0^3 S_2^3)$$
$$+ \alpha^j(S_0^2 S_1 S_2^3 + S_0^3 S_2^2 S_3 + S_1^5 S_2 + S_0 S_1^4 S_3)$$
$$+ (S_0^2 S_2^4 + S_0^2 S_1 S_2^2 S_3 + S_1^4 S_2^2 + S_1^5 S_3) = 0 \qquad (7.25)$$

Notice that the cubic term (α^{3j}) has gone, so we expect only two solutions now. Substituting our syndromes into (7.25) yields

$$\alpha^{2j+5} + \alpha^{j+6} + \alpha = 0$$

This time only the values $j = 3$ and $j = 7$ (or 0) are satisfied. Since we introduced the errors at positions 3 and 7, these results are correct. To calculate the error patterns we simply have to substitute i and j into (7.19). Let $j = 3$ and $i = 7$.

$$S_1 = (S_0 + \varepsilon_j)\alpha^i + \varepsilon_j\alpha^j$$

$$\varepsilon_3 = (S_1 + S_0\alpha^i)/(\alpha^i + \alpha^j)$$

$$\varepsilon_3 = (\alpha^5 + \alpha^6)/(\alpha^3 + \alpha^7)$$

$$\mathbf{\varepsilon_3 = (111 + 101)/(011 + 001) = \alpha^7 \text{ or } 001}$$

Lastly, from (7.15)

$$S_0 = \varepsilon_i + \varepsilon_j$$

$$\mathbf{\varepsilon_7 = \alpha^6 + \alpha^7 = 101 + 001 = 100}$$

Table 7.5 Evaluating the syndromes for a single error

Evaluating S_0			Evaluating S_1			Evaluating S_2			Evaluating S_3		
A	α^5	111	$A\alpha^7$	α^5	111	$A\alpha^7$	α^5	111	$A\alpha^7$	α^5	111
B	α^6	101	$B\alpha^6$	α^5	111	$B\alpha^5$	α^4	110	$B\alpha^4$	α^3	011
C	α^5	111	$C\alpha^5$	α^3	011	$C\alpha^3$	α^1	010	$C\alpha^1$	α^6	101
P	α^1	010	$P\alpha^4$	α^5	111	$P\alpha^1$	α^2	100	$P\alpha^5$	α^6	101
Q	α^6	101	$Q\alpha^3$	α^1	100	$Q\alpha^6$	α^5	111	$Q\alpha^2$	α^1	010
R	α^6	101	$R\alpha^2$	α^1	010	$R\alpha^4$	α^3	011	$R\alpha^6$	α^5	111
S	α^3	011	$S\alpha^1$	α^4	110	$S\alpha^2$	α^5	111	$S\alpha^3$	α^6	101
S_0		100	S_1		100	S_2		100	S_3		100

So we have located and calculated both the errors introduced into the codeword. To correct the errors, A' is eXORed with ε_7 while Q' is eXORed with ε_3. What happens, however, if there is only a single error? Let's try again but without ε_3. Recalculating the syndromes gives the results shown in Table 7.5.

$$S_0 = \alpha^2, \; S_1 = \alpha^2, \; S_2 = \alpha^2 \text{ and } S_3 = \alpha^2$$

Substituting these syndromes into (7.25) gives

$$\alpha^{2j}(\alpha^5 + \alpha^5 + \alpha^5 + \alpha^5) + \alpha^j(\alpha^5 + \alpha^5 + \alpha^5 + \alpha^5) + \alpha^5 + \alpha^5 + \alpha^5 + \alpha^5 = 0$$

$$0 = 0$$

This result alerts us to the fact that there are not two errors. Since the syndromes are non-zero we may assume that there is a single error at i, so using (for one example) (7.16) and (7.17)

$$\alpha^i = (S_2)/(S_1) = 1 \text{ or } \alpha^7 \text{ so } i = 7$$

i could just as readily have been found from S_1/S_0, or S_3/S_2, the choice was quite arbitrary, and we could use this fact to verify that there was only a single error.

Since we know that there is only one error

$$S_0 = \varepsilon_i + \varepsilon_j \text{ and } \varepsilon_j = 0 \text{ so } \varepsilon_7 = S_0 \text{ or } 100$$

7.4 DISCUSSION

Comparing the single error-correcting solution with the double-correcting solution it can be seen that there is an exponential increase in the work involved, both in terms of deriving a solution and locating and correcting the errors. This has serious implications for the hardware design as the level of error protection required has a radical effect on the implementation. In other words, the design of an error-correction system for two errors, for example, is unique. Ideally it is preferable to have a general-purpose solution where the required level of protection is easily set, i.e. a standard processing architecture that is appropriate to a wide range of error-correction capability.

Later we'll examine techniques for creating a slightly more general-purpose solution (and much less work) for evaluating the time domain check symbols. The technique is like erasure, where the check symbols are set to zero and syndromes are found, much as would be done at the receiver. Working back from the syndromes to our known 'errors', the check symbols, is more straightforward than the kind of symbol manipulation given in this chapter.

You should at this point be able to construct the equations necessary to give an arbitrary level of error protection to a message. Working these through to single expressions for evaluating the check symbol values will require a little practice and is, as it turns out, unnecessary. In short, this example is not proposed as a serious solution for a double-symbol correcting scheme (although it is quite functional), but has been included to illustrate the complexities of coding in this way.

Part Four

Frequency domain Reed–Solomon coding

8

Frequency domain coding

Finite fields are amenable to Fourier transform-like operations which move the data through orthogonal domains in a very similar manner to which the Fourier transform provides a doorway between time and frequency. Just as the Fourier transform extracts pure sine waves from time domain data, the equivalent operation on a finite field extracts features from the time domain data and makes them visible to the observer, although these features are not of course sine waves.

8.1 CALCULATING A FOURIER TRANSFORM

The Fourier transform over GF(2³) takes the form

$$F_i = \sum_{j=0}^{6} d_j \alpha^{ij}$$

There is a slight modification of the basic numbering strategy used earlier. Here the element index i starts at the left as 0, and increases to the right up to 6. Over seven data elements d_0 to d_6, the Fourier transform is thus

$$F_0 = d_0\alpha^0 + d_1\alpha^0 + d_2\alpha^0 + d_3\alpha^0 + d_4\alpha^0 + d_5\alpha^0 + d_6\alpha^0$$
$$F_1 = d_0\alpha^0 + d_1\alpha^1 + d_2\alpha^2 + d_3\alpha^3 + d_4\alpha^4 + d_5\alpha^5 + d_6\alpha^6$$
$$F_2 = d_0\alpha^0 + d_1\alpha^2 + d_2\alpha^4 + d_3\alpha^6 + d_4\alpha^1 + d_5\alpha^3 + d_6\alpha^5$$
$$F_3 = d_0\alpha^0 + d_1\alpha^3 + d_2\alpha^6 + d_3\alpha^2 + d_4\alpha^5 + d_5\alpha^1 + d_6\alpha^4$$
$$F_4 = d_0\alpha^0 + d_1\alpha^4 + d_2\alpha^1 + d_3\alpha^5 + d_4\alpha^2 + d_5\alpha^6 + d_6\alpha^3$$
$$F_5 = d_0\alpha^0 + d_1\alpha^5 + d_2\alpha^3 + d_3\alpha^1 + d_4\alpha^6 + d_5\alpha^4 + d_6\alpha^2$$
$$F_6 = d_0\alpha^0 + d_1\alpha^6 + d_2\alpha^5 + d_3\alpha^4 + d_4\alpha^3 + d_5\alpha^2 + d_6\alpha^1$$

where $\alpha^0 = \alpha^7 = 1$.

To help distinguish between time and frequency domain values, time domain variables will be lower case while frequency domain variables will be upper case. So what does this transform tell us? If we transform the data sequence calculated for the single error-correcting example previously in Chapter 6, we get the result

Time domain (original)

$$= 6, 0, 2, 4, 7, 0, 7$$
$$= \alpha^4, 0, \alpha^1, \alpha^2, \alpha^5, 0, \alpha^5$$

Frequency domain

$$= 0, 0, 5, 6, 0, 4, 1$$
$$= 0, 0, \alpha^6, \alpha^4, 0, \alpha^2, \alpha^7$$

again, but using the data from the double-correcting example of Chapter 7:

Time domain (original)

$$= 3, 5, 7, 2, 5, 5, 3$$
$$= \alpha^3, \alpha^6, \alpha^5, \alpha^1, \alpha^6, \alpha^6, \alpha^3$$

Frequency domain

$$= 0, 4, 3, 4, 0, 0, 0$$
$$= 0, \alpha^2, \alpha^3, \alpha^2, 0, 0, 0$$

In the first case the frequency domain contains two zeros followed by five symbols (of which incidentally one is also zero), while in the second case there are three non-zero elements. In the first case the first two zeros are in positions F_0 and F_1. The equations used to derive these two frequency elements are identical to (6.1) and (6.2) which were the starting conditions of the single error-correcting code. Since we had arranged for these to be zero when calculating R and S, the result here is not surprising. The third zero at position 4, is an 'accident' of the data, but tells us that there is some extra redundancy in our data which we may be able to exploit if we so choose. This, however, is not the subject in hand. Looking to the second example where we have transformed the double correcting message, the four zeros correspond to four equations which are identical to (7.1) to (7.4), the starting conditions used to set the redundant symbols. Where we see a zero in the frequency domain the polynomial is said to have a root in the time domain. Superficially the Fourier transform has shown us how much data is present in the encoded message. R and S in the first example, and P, Q, R and S in the second example, are calculated solely from the data provided and thus convey no **extra** information. They repeat in a fashion, or reinforce the data. The Fourier transform is simply telling us this. A very simple comparison could be drawn with a sine wave in the time domain. If we add extra points into the time domain signal which interpolate or sit on the sine wave, then the Fourier transform will not reveal any extra content in the signal in the frequency domain, but will show an improvement in the signal-to-noise ratio – a reinforcing.

Because the time and frequency domains are orthogonal we can move freely back and forth between them. The inverse transform over the finite field GF(2^3) is given by

$$d_i = \sum_{j=0}^{6} F_j \alpha^{-ij}$$

So how does the frequency domain help? If by encoding redundancy into the message we add two zeros per protectable error into the frequency domain description of the message, the corollary is that we could start in the frequency domain, append $2t$ zeros to our data and perform an inverse Fourier transform. This will provide us with an encoded time domain message that has all the properties of redundancy we desire, but without any of the error-specific calculations that were previously necessary. In other words, to set the level of error protection all we need to do is to add the right number of zeros before converting to the time domain.

The actual message that is transmitted looks nothing like the original data and more work will be required to extract it. The constant and very structured nature of the transformation, however, will often more than offset this in terms of hardware.

8.2 USING THE FREQUENCY DOMAIN TO ENCODE A MESSAGE

Upon receipt of the time domain message the receiver must perform a Fourier transform to recover the original data which in this regime started life in the frequency domain. In the event of no errors the $2t$ message positions that were preloaded with zeros will still be zero. Suppose we start with a double-symbol correcting message

$$0, 0, 0, 0, 3, 5, 4$$

The transmitted time domain message is

$$2, 4, 1, 7, 6, 3, 5$$

From this the data are quite invisible. Fourier transforming this yields our original message:

$$0, 0, 0, 0, 3, 5, 4$$

If the message gets corrupted by errors within the code's capacity to correct, then the zeros will become non-zero values. For example, suppose the message

$$2, 4, 1, 7, 6, 3, 5$$

is corrupted by the double-symbol error

$$3, 0, 4, 0, 0, 0, 0$$

The received message will be

$$\textbf{1}, 4, \textbf{5}, 7, 6, 3, 5$$

and after Fourier transforming

$$\textit{7, 5, 6, 1, }3, 1, 6$$

Since the first four symbols (italic) contain non-zero values, we are alerted to the presence of an error(s). The last three symbols are a combination of both data and error and since we know neither, they are discarded. Because of the orthogonal nature of the two domains, however, in spite of errors the data can have no impact on the first four symbols which represent errors only. This can be easily demonstrated by performing a Fourier transform on the error vector, a luxury that is not normally available. The frequency domain form of the error vector is

$$7, 5, 6, 1, 0, 4, 2$$

The first four symbols are identical to the first four of the corrupted message. The problem is, how do we get back from the four known frequency domain error symbols to the complete error vector? Knowing the form of the Fourier transform we could generate four simultaneous equations based upon the four known frequency error spectra, i.e.

$$F_0 = \varepsilon_i \alpha^{0i} + \varepsilon_j \alpha^{0j} = 7 \text{ (or } \alpha^5)$$
$$F_1 = \varepsilon_i \alpha^{1i} + \varepsilon_j \alpha^{1j} = 5 \text{ (or } \alpha^6)$$
$$F_2 = \varepsilon_i \alpha^{2i} + \varepsilon_j \alpha^{2j} = 6 \text{ (or } \alpha^4)$$
$$F_3 = \varepsilon_i \alpha^{3i} + \varepsilon_j \alpha^{3j} = 1 \text{ (or } \alpha^7)$$

These equations are of course identical to the syndromes (7.15) to (7.18) used to solve the double errors in the time domain example. Solution of these equations poses us much the same kinds of problems of generality faced earlier, and is not the preferred approach. With two errors we have four unknowns, two positions and two patterns. Knowing four of the frequency domain elements means that there is enough information to calculate the errors and their positions, demonstrated by the four simultaneous equations above. From a philosophical point of view what this says is that the four elements F_0 to F_3 are unique to this particular error. Furthermore, the three unknown elements F_4 to F_6 are also unique to these four elements. Because the error vector has at least five zeros, the transformed equivalent will have at least five roots. In short, there is only one polynomial that contains as least five roots and has the four elements F_0 to F_3. Knowing this we could find the complete error vector using a look-up table but this would get rather large for realistic fields (i.e. with more than three bits per element).

8.3 CORRECTING ERRORS IN THE FREQUENCY DOMAIN

The solution to the problem of locating and identifying the errors lies in the use of a **recursive extension circuit** (REC) to complete the error spectrum. Once complete the spectrum may be returned to the time domain to yield the actual error. The circuit, by means of feedback, is able to generate polynomials with set numbers of roots corresponding to the number of feedback elements. The known elements of the error spectrum are used to set up the starting conditions of the recursive extension circuit which then free-runs to generate the remaining unknown elements. The simplest REC will generate the most predictable polynomial, i.e. with the most roots. An example of such a circuit is given in Fig. 8.1.

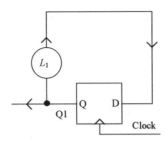

Fig 8.1 Generating a vector with six roots.

If a vector or polynomial has six roots, then the transformed equivalent will have six zeros and one non-zero element over GF(2^3). The position of the non-zero element is fixed by the feedback multiplier L_1, while the value of the non-zero element is set by the initial value in the register. Suppose we start with $L_1 = \alpha^2$ and preload the register with 6 or α^4. The output from the circuit will be α^{2i+4} where i is the clock count or symbol position. For $L_1 = \alpha^2$ and a starting value of α^4 the generated sequence is as shown in Table 8.1.

Table 8.1 Output from a simple REC

Clocks (i)	Calculation	Output	Decimal value
0	α^4	α^4	6
1	$\alpha^4.\alpha^2$	α^6	5
2	$\alpha^6.\alpha^2$	α^1	2
3	$\alpha^1.\alpha^2$	α^3	3
4	$\alpha^3.\alpha^2$	α^5	7
5	$\alpha^5.\alpha^2$	α^7	1
6	$\alpha^7.\alpha^2$	α^2	4
7	$\alpha^2.\alpha^2$	α^4	6 *Repeats*
8	$\alpha^4.\alpha^2$	α^6	5
	etc.		

Performing an inverse Fourier transform on the vector

$$6, 5, 2, 3, 7, 1, 4$$

gives

$$0, 0, 6, 0, 0, 0, 0$$

The non-zero element is 6 (the initial contents of the register), and it is in position 2 (corresponding to the feedback multiplier α^2). If we know any two adjacent values in the spectrum, then we can complete it using the REC. Take, for example, the single symbol-correcting example of Chapter 6. The original message was

$$6, 0, 2, 4, 7, 0, 7$$

and it was earlier shown that the Fourier transform of this message is

$$0, 0, 5, 6, 0, 4, 1$$

In Section 6.4, the message was corrupted by a single-symbol error

$$0, 0, 0, 0, 3, 0, 0$$

such that it appeared as

$$6, 0, 2, 4, \mathbf{4}, 0, 7$$

and the Fourier transform of this is

$$3, 1, 3, 4, 7, 0, 4$$

We know that the first two elements (3 and 1), represent error only since they were 0 in the uncorrupted and Fourier transformed message. Using these, we know that

$$1 = L_1.3 \text{ or } \alpha^7 = L_1.\alpha^3$$

so

$$L_1 = \alpha^4$$

Repeating Table 8.1, but with $L_1 = \alpha^4$ and the register preloaded with 3, we get the sequence shown in Table 8.2. This tells us that the complete spectrum for the added error is

$$3, 1 \text{ (known)}, 6, 2, 7, 4, 5$$

The inverse Fourier transform of this completed spectrum is

$$0, 0, 0, 0, 3, 0, 0$$

the added error. So the REC has reconstructed the error pattern from the two known elements of the error frequency spectrum.

Table 8.2 Finding an error using the REC

Clocks (i)	Calculation	Output	Decimal value
0	α^3	α^3	3
1	$\alpha^3.\alpha^4$	α^7	1
2	$\alpha^7.\alpha^4$	α^4	6
3	$\alpha^4.\alpha^4$	α^1	2
4	$\alpha^1.\alpha^4$	α^5	7
5	$\alpha^5.\alpha^4$	α^2	4
6	$\alpha^2.\alpha^4$	α^6	5
7	$\alpha^6.\alpha^4$	α^3	3 *Repeats*
8	$\alpha^3.\alpha^4$	α^7	1
	etc.		

Extending the idea to double errors, the circuit must generate an error spectrum with five roots corresponding to the five time domain zeros. Such a circuit is shown in Fig. 8.2.

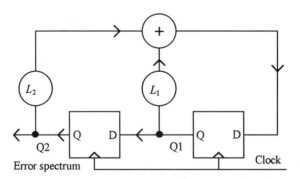

Fig. 8.2 REC for generating a vector with five roots.

Again the feedback elements are a function solely of the positions of the two non-zero time domain elements. If the errors are at positions i and j, then the value of L_2 is given by α^{i+j} while L_1 is given by $\alpha^i + \alpha^j$. You may notice (remembering Section 6.1.6) that L_1 and L_2 form a quadratic of the form

$$\alpha^{2x} + \alpha^x L_1 + L_2 = 0$$

where x has two roots i and j. This can be demonstrated in the following way. Suppose that we have set up a double-correcting message where after errors we have four known error spectra E_0 to E_3. Whatever else happens, we must at least ensure that the circuit outputs these four. By preloading the registers with E_0 and E_1, we guarantee that they will emerge. Now we must arrange for E_2 to appear on the input to the right-hand register, ready to be latched in next and hence follow E_1 out of the circuit, so we must satisfy

$$E_0 L_2 + E_1 L_1 = E_2$$

From the Fourier transform, if the errors are e_i and e_j, at positions i and j, then

$$E_0 = e_i + e_j, \; E_1 = e_i \alpha^i + e_j \alpha^j, \; E_2 = e_i \alpha^{2i} + e_j \alpha^{2j}$$

Substituting these into the above yields

$$e_i L_2 + e_j L_2 + e_i \alpha^i L_1 + e_j \alpha^j L_1 = e_i \alpha^{2i} + e_j \alpha^{2j}$$

Since e_i and e_j are independent, we can split this into two parts to give

$$e_i L_2 + e_i \alpha^i L_1 = e_i \alpha^{2i} \quad \text{and} \quad e_j L_2 + e_j \alpha^j L_1 = e_j \alpha^{2j}$$

Clearly the errors e_i and e_j can be cancelled from both sides of the equations to give the solutions

$$L_2 = \alpha^{i+j} \quad \text{and} \quad L_1 = \alpha^i + \alpha^j$$

Previously the error vector

$$3, 0, 4, 0, 0, 0, 0$$

was used to demonstrate the orthogonal relationship between the time and frequency domain. This vector has errors at positions 0 and 2 and using these we would expect $L_2 = \alpha^2$ and $L_1 = \alpha^6$. Normally we will not have this information and so will have to compute them from the four known error spectra as follows. From this previous example, we have part of an error spectrum 7, 5, 6, 1, X, X, X, where X is unknown. To set up the circuit, the two registers are first preloaded such that $Q_2 = 7$ and $Q_1 = 5$. This way, when the circuit is clocked it will definitely output the first two values of the sequence. In order for the REC to output the next value, 6, we must arrange for

$$7.L_2 + 5.L_1 = 6$$

By so doing, 6 will be clocked into the right-hand register after the first clock, and we must arrange for the final known value, 1, to be generated using

$$5.L_2 + 6.L_1 = 1$$

Solving these we see that indeed $L_2 = \alpha^2$ and $L_1 = \alpha^6$ as already predicted. Table 8.3 below shows the results of clocking the circuit six times. The complete error spectrum appears in column Q_2.

Table 8.3 Completing the error spectrum for two errors

Q_2	Q_1	$\alpha^2 Q_2$	$\alpha^6 Q_1$	D_1	D_1 decimal
7	5	α^7	α^5	α^4	6
5	6	α^1	α^3	α^7	1
6	1	α^6	α^6	0	0
1	0	α^2	0	α^2	4
0	4	0	α^1	α^1	2
4	2	α^4	α^7	α^5	7
2	7	α^3	α^4	α^6	5

Having completed the spectrum, which is

$$7, 5, 6, 1, 0, 4, 2$$

we can inverse Fourier transform this to yield the time domain error vector,

$$3, 0, 4, 0, 0, 0, 0$$

Once we have this then it can be XORed with the received message to give a corrected time domain message. The time domain message is then Fourier transformed to return the original data plus zeros. It should be noted that in order for this to work, the zeros in the original message must be adjacent. This is not a requirement of the basic frequency domain technique, but of the use of recursive extension in the decoder. The idea of recursive extension can be compared with a quadratic curve of the form

$$ax^2 + bx + c = y$$

for example. Once we know three points on the curve, the rest of the curve is known. The REC simply provides a convenient mechanism for achieving something similar over finite fields.

To extend the REC to error spectra with fewer roots, more registers and feedback are added. As a rule, one register and one feedback element are required per symbol error that is to be corrected. It may appear that we have moved away from a general solution to correction, to one which is specific to the number of errors that are to be corrected. The REC is easily scaled to any number of errors by making it large enough to accommodate the most errors that will be handled and setting the feedback to zero where fewer will be corrected. A more pressing problem is that of solving the attendant simultaneous equations in order to find the values of the feedback multipliers. The thing to note, however, is that the solution of L_1 and L_2 is relatively simple when compared with working backwards from the FFT using the four known error spectra. The REC itself may be easily generalized but the calculation of the feedback multipliers must be done via a set of standard solutions or by using a Gaussian elimination or matrix inversion-type operation.

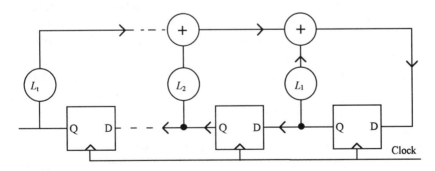

Fig. 8.3 Extending the REC for more errors.

The standard solution for a single error-correcting system is

$$L_1 = E_1/E_0 \text{ (all others set to zero)}$$

where E_0 and E_1 are the two adjacent frequency components that should have been zero. For a double-correcting code the solution is

$$L_1 = (E_0.E_3 + E_1.E_2)/(E_1.E_1 + E_0.E_2)$$

and

$$L_2 = (E_1.E_3 + E_2.E_2)/(E_1.E_1 + E_0.E_2)$$

Appendix B gives a list of solutions for a variety of levels of error protection. As with time domain processing, if we know the positions of the errors in advance, we can correct more errors. If for example we have set up a single error-correcting code but subsequently find that two errors have occurred, one at symbol position i and one at symbol position j, then using $L_2 = \alpha^{i+j}$ and $L_1 = \alpha^i + \alpha^j$, we can set up a double-correcting REC. Since we already know L_1 and L_2 we need only preload the circuit with two error spectra (which is all we know for a single correcting code). The REC will then provide a complete error spectrum for the two errors allowing us to correct them both.

8.4 HOW DOES RECURSIVE EXTENSION WORK?

While it is by no means essential, you may be interested to know how the REC works. Examination of the generalized circuit in Fig. 8.3 above shows that it performs a 'sum of products' type operation. There is a vector (part of the error spectrum) contained in the registers and each vector element is multiplied by a weight and summed to produce a value which forms the next input to the registers. This operation is also called a **convolution** and its length is a function of the number of errors that are to be corrected.

If you are familiar with signal processing you will have heard the expression '*multiplication in the frequency domain is the same as convolution in the time domain*' or, indeed, '*multiplication in the time domain is the same as convolution in the frequency domain*'. This rule happens to be true over finite fields and if we have an error vector over $GF(2^3)$, for example, we can invent another vector which, when multiplied by the error, will produce 0.

Using our previous error

$$e = 3, 0, 4, 0, 0, 0, 0$$

we could create an **error locator polynomial**

$$l = 0, l_1, 0, l_3, l_4, l_5, l_6$$

Regardless of the values of l_i, the result will be zero if we multiply the two vectors together. Equally, if we Fourier transform l, to give L in the frequency domain, and then convolve L with E, the error spectrum, the result will also be zero. The convolution has the form

$$\sum_{j=0}^{t} L_j E_{k-j} = 0$$

You may notice that the convolution takes place over $t + 1$ elements of the spectrum, while the REC performs a convolution over only t elements. In fact, the REC has an invisible element which is shown in Fig. 8.4 below.

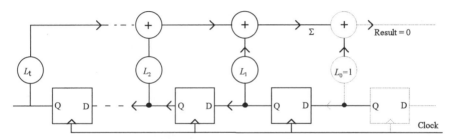

Fig. 8.4 Convolution using the REC.

If the registers are loaded sequentially with $t + 1$ elements from the error spectrum, E_k to E_{k-t} (k arbitrarily setting the window of $t + 1$ errors over the spectrum) then the summation output from the circuit must be 0. The choice of l_i is arbitrary so we can equally arrange for at least one value of L_j to be arbitrary and the one we choose is L_0 which is set to 1. Since this is the case, for the result of the convolution to be zero the value in the dotted register must be equal to the sum of products from all the other elements (Σ). In other words, Σ is the same value that is in the invisible register. Because this is so we don't need the right-hand dotted register since we can feed Σ back into the right-hand solid register to give the circuit of Fig. 8.3.

We can perform the convolution longhand to demonstrate that this is in fact what is going on. Using our double error-correcting example we have L_0 to L_2, so we can form two convolutions over the four known error spectra:

$$L_0.E_3 + L_1.E_2 + L_2.E_1 = 0$$

and

$$L_0.E_2 + L_1.E_1 + L_2.E_0 = 0$$

Now L_0 is set to 1 so these become

$$E_3 + L_1.E_2 + L_2.E_1 = 0$$

and

$$E_2 + L_1.E_1 + L_2.E_0 = 0$$

These are sometimes called the **key equations**. Solving gives

$$L_1 = (E_0.E_3 + E_1.E_2)/(E_1.E_1 + E_0.E_2)$$

and

$$L_2 = (E_1.E_3 + E_2.E_2)/(E_1.E_1 + E_0.E_2)$$

This is of course the same solution we obtained previously, and if we put in the four known error spectra we get the following

$$E_0 = 7 \ (\alpha^5), \ E_1 = 5 \ (\alpha^6), \ E_2 = 6 \ (\alpha^4), \ E_3 = 1 \ (\alpha^7)$$

$$L_1 = (\alpha^5.\alpha^7 + \alpha^6.\alpha^4)/(\alpha^6.\alpha^6 + \alpha^5.\alpha^4) = \alpha^6$$

$$L_2 = (\alpha^6.\alpha^7 + \alpha^4.\alpha^4)/(\alpha^6.\alpha^6 + \alpha^5.\alpha^4) = \alpha^2$$

which is the expected answer. It should be clear now why the weights L_1 to L_t depend only on the positions of the errors and not on their patterns, since they are in fact a function of the error locator polynomial and hence the error location (remember that the locator l is 0 at the error positions). We can demonstrate this by constructing l. In the previous example, we were looking for two errors and we knew L_1 and L_2. It has already been demonstrated that L_1 and L_2 lead to the quadratic

$$\alpha^{2x} + \alpha^x L_1 + L_2 = 0$$

for an REC capable of completing an error spectrum with two errors. Using this we can construct l by cycling x through 0 to 6 and noting the result.

This is shown in Table 8.4.

Table 8.4 Locating errors using the REC

x	Quadratic	Result	Decimal	=0?
0	$\alpha^0 + \alpha^0\alpha^6 + \alpha^2$	0	0	✓
1	$\alpha^2 + \alpha^1\alpha^6 + \alpha^2$	α^7	1	✗
2	$\alpha^4 + \alpha^2\alpha^6 + \alpha^2$	0	0	✓
3	$\alpha^6 + \alpha^3\alpha^6 + \alpha^2$	α^6	5	✗
4	$\alpha^1 + \alpha^4\alpha^6 + \alpha^2$	α^6	5	✗
5	$\alpha^3 + \alpha^5\alpha^6 + \alpha^2$	α^7	1	✗
6	$\alpha^5 + \alpha^6\alpha^6 + \alpha^2$	α^2	4	✗

From Table 8.4, we can see that the quadratic is satisfied at the error positions. l is thus

$$0, 1, 0, 5, 5, 1, 4$$

This won't tell us what the errors are, simply where, but this is sometimes useful, and a lot less computationally expensive. As such, this result can be used as a short cut to the time domain error vector. Previously, having found L_1 and L_2 the REC was clocked to complete the error spectrum and the spectrum inverse Fourier transformed to yield the original errors. Instead, by completing l (which could be found analytically in the quadratic case) and which takes no longer than completing the error spectrum by recursive extension, we can locate the errors. From the Fourier transform we know that

$$F_0 = \varepsilon_i + \varepsilon_j$$

and

$$F_1 = \alpha^i\varepsilon_i + \alpha^j\varepsilon_j$$

We know i and j from $l_0 = 0$ and $l_2 = 0$, so rearranging the above leads to

$$\varepsilon_i = (F_1 + F_0\alpha^j)/(\alpha^i + \alpha^j)$$

and

$$\varepsilon_j = \varepsilon_i + F_0$$

so we now have the patterns as well. In this example, F_0 was 7 or α^5 and F_1 was 5 or α^6. Substituting them in gives

$$\varepsilon_0 = (\alpha^6 + \alpha^5\alpha^2)/(\alpha^0 + \alpha^2)$$

$$\varepsilon_0 = \alpha^2/\alpha^6 = \alpha^3 \text{ or } 3$$

and

$$\varepsilon_2 = \alpha^3 + \alpha^5 = \alpha^2 \text{ or } 4$$

which are the correct answers.

8.5 CORRECTION LIMITS

We have now seen two methods of locating and correcting errors and it is important that we understand the limits of the correction codes. For every $2t$ added zeros in the message, t symbol errors may be corrected. The solutions that have been presented, however, are valid only where the number of actual errors equals the number solved for. In other words, if we set up a message with four zeros and solve for two errors, if only one error has actually occurred then the final solution will be incorrect. In the event of non-zero syndromes in a decoded message, the simplest approach to this problem is to solve for n errors, starting at $n = 1$, increasing to t, until the number of errors (non-zero elements) in the reconstructed time domain error vector equals n. Where there is a mismatch between n and the actual number of errors, the time domain error vector takes on a random appearance usually containing many more than t errors. This clearly identifies it as incorrect. Using a matrix inversion to solve the key equations will result in an all-zero solution if fewer errors exist than are solved for.

The alternative to trial and error is to find the order of the locator polynomial used in the recursive extension process, i.e. how many registers and feedback multipliers will be required. This information is contained in the syndromes of the received and decoded message. Unfortunately, finding the polynomial order amounts to much the same task as attempting a correction since testing for power requires solving (or part solving), once again, the key equations. If matrix inversion is used to solve the key equations (examples of which may be found in Appendix E), then small savings can be made by noting that we need solve only one row of the inverse matrix in order to determine whether or not the order of the proposed solution matches the number of errors present.

In a previous section we saw how the frequency domain error syndromes were used to calculate the feedback multipliers in the recursive extension circuit. If we know eight syndromes and are solving for four errors then we would have four equations:

$$L_4E_0 + L_3E_1 + L_2E_2 + L_1E_3 = E_4$$
$$L_4E_1 + L_3E_2 + L_2E_3 + L_1E_4 = E_5$$
$$L_4E_2 + L_3E_3 + L_2E_4 + L_1E_5 = E_6$$
$$L_4E_3 + L_3E_4 + L_2E_5 + L_1E_6 = E_7$$

where L are feedback multipliers and E are the known error spectra. These can be rearranged into a matrix form as:

$$\begin{bmatrix} E_0 & E_1 & E_2 & E_3 \\ E_1 & E_2 & E_3 & E_4 \\ E_2 & E_3 & E_4 & E_5 \\ E_3 & E_4 & E_5 & E_6 \end{bmatrix} \begin{bmatrix} L_4 \\ L_3 \\ L_2 \\ L_1 \end{bmatrix} = \begin{bmatrix} E_4 \\ E_5 \\ E_6 \\ E_7 \end{bmatrix}$$

or

$$E.L = R$$

To evaluate the matrix L, we must invert E so

$$L = R.E^{-1} \text{ and } E.E^{-1} = 1$$

If there are fewer than four errors present, then the last identity, $E.E^{-1} = 1$, will not hold true. In fact, E^{-1} will contain all zeros. To test this identity only one row or column of E^{-1} need be evaluated which represents a saving over evaluating the entire inversion. In some later examples we'll see in more detail the effects of a mismatch between actual errors and errors solved for.

8.6 SUMMARY OF FREQUENCY DOMAIN CODING

1. To summarize frequency domain error control coding, we decide the field over which we will operate, and
2. generate a data set comprising $2t$ leading zeros and 2^m-1-2t data symbols, where the field is $GF(2^m)$. Each symbol will be m bits long.
3. The resulting frequency domain data set is inverse Fourier transformed to give a time domain message which is transmitted.

4.   At the receiver the incoming message is forward Fourier transformed in an attempt to reconstruct the original frequency domain message.
5.   In the event of no detected errors, the leading $2t$ zeros will be present and the data are recovered.
6.   If some or all of the $2t$ zero positions contain non-zero symbols then errors have occurred and the $2t$ error spectrum elements must be used to solve the key equations in order to construct the recursive extension circuit. This will entail establishing how many errors have occurred.
7.   The REC is clocked to complete the error spectrum and the result is inverse Fourier transformed to form the time domain error pattern.
8.   The time domain error pattern is XORed with the received, corrupted message to correct it.
9.   The corrected message is forward Fourier transformed to reconstruct the original frequency domain message.

This sequence is shown in Fig. 8.5.

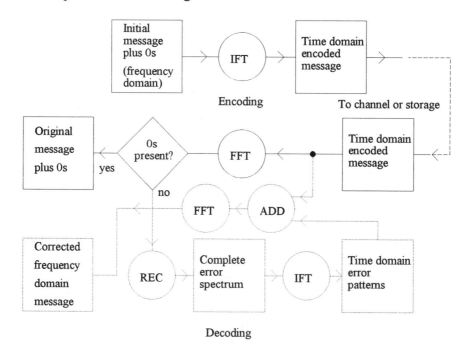

Fig. 8.5 The frequency domain encoding/decoding route.

In Fig. 8.5, the correctional part of the system is shown dotted. The only slightly thorny issue which is naturally of great interest to the hardware designer, is the solution of the key equations. This may be done via a table of standard solutions for the necessary levels of protection that are to be provided, or a more general elimination or matrix inversion-type solution may be sought. If a micro-controller is present in the hardware then the latter is simple enough to achieve, although by using multi-dimensional finite fields (Chapter 10), large blocks may be economically encoded using small fields, making the solutions somewhat easier to obtain.

8.7 AN EXAMPLE OVER GF(2⁸)

To provide a realistic example of error coding in the frequency domain we'll consider a message over $GF(2^8)$. The block that we must construct has 255 symbols each 8 bits, making a total block size of 2040 bits which is quite reasonable on, say, a packet network. In this example we'll guard against 8 errors (hence $t = 8$) so we'll need 16 zeros placed with the data. Using a piece of text from this book, encoded as ASCII, and the primitive polynomial 285, the message appears as follows in Table 8.5.

Table 8.5 Original frequency domain data

0	*0*	*0*	*0*	*0*	*0*	*0*	*0*	*0*	*0*	*0*	*0*	*0*	*0*	*0*	*0*
I	f		t	h	e		r	e	g	i	s	t	e	r	s
	a	r	e		p	r	e	-	l	o	a	d	e	d	
w	i	t	h		a		n	o	n	-	z	e	r	o	
v	a	l	u	e	,		t	h	e	n		s	u	c	c
e	s	s	i	v	e		c	l	o	c	k	s		w	i
l	l		p	r	o	d	u	c	e		s	u	c	c	e
s	s	i	v	e		e	l	e	m	e	n	t	s		i
n		t	h	e		f	i	e	l	d	.		A	s	
f	a	r		a	s		m	a	n	i	p	u	l	a	t
i	n	g		t	h	e		e	l	e	m	e	n	t	s
	i	s		c	o	n	c	e	r	n	e	d	,		a
	c	h	o	i	c	e		m	u	s	t		b	e	
m	a	d	e		b	e	t	w	e	e	n		r	e	p
r	e	s	e	n	t	i	n	g		t	h	e		e	l
e	m	e	n	t	s		a	s		t	h	e	i	r	

Table 8.6 below shows the message after inverse Fourier transforming into the time domain. This is the message that will actually be transmitted. The data are shown in a matrix format only for the purpose of presentation, and their values are expressed in hexadecimal.

Table 8.6 Time domain message after inverse Fourier transforming Table 8.5

70	9A	92	C0	CF	23	A4	F0	3D	99	8A	88	2D	52	6F	29
68	A6	19	D4	D6	32	69	A9	44	E5	21	8D	FB	9C	07	6C
9D	88	70	4C	FF	51	53	F2	5E	5B	F7	17	F6	FC	29	42
B3	23	C0	1A	E5	9D	EA	42	76	C0	62	3C	6A	C1	9F	C6
BB	F5	7A	BE	75	CA	C0	5B	CD	09	8E	8B	7A	78	78	2D
89	18	ED	95	4C	B8	CE	48	52	BA	B9	66	B7	6D	55	E7
7A	80	DF	B4	A4	3D	72	E7	B9	61	29	DE	C4	F7	61	22
A5	C5	EC	24	E9	78	72	AE	70	88	C3	E6	63	2D	1E	B8
CA	41	60	49	96	3C	5D	28	FB	BF	04	62	EB	AD	7F	9F
EE	AD	F0	DF	D4	75	3D	AC	91	AC	8E	F8	AC	1C	4B	4D
3C	78	F8	A5	43	58	52	0B	C6	A7	D0	44	94	CE	2C	4E
1F	2E	78	40	60	6A	DB	F9	27	5E	F7	61	FF	C1	3F	CC
A6	26	9B	0E	17	B2	E9	A4	4A	2D	84	59	68	7F	F3	D6
DC	C1	98	6D	33	E2	A1	12	16	20	95	DB	8F	A8	44	41
E3	6E	17	C9	58	ED	6E	AC	68	4B	2F	09	6C	62	03	00
76	C9	3F	14	42	5E	73	73	B4	EA	8D	FF	46	CA	08	

Table 8.7 The received (corrupted) message

70	9A	92	C0	CF	23	A4	F0	3D	99	8A	88	2D	52	6F	29
68	A6	19	D4	D6	32	69	A9	44	E5	21	8D	FB	9C	07	6C
9D	88	70	4C	FF	51	53	F2	5E	5B	F7	17	F6	FC	29	42
B3	23	C0	1A	E5	9D	EA	42	76	C0	62	3C	6A	C1	9F	C6
BB	F5	7A	BE	75	CA	**A1**	5B	CD	09	8E	8B	7A	78	78	2D
89	18	ED	95	4C	B8	CE	48	52	BA	B9	66	B7	6D	55	E7
7A	80	DF	B4	A4	3D	72	E7	B9	61	29	DE	C4	F7	61	22
A5	C5	EC	24	E9	78	72	AE	**F4**	88	C3	E6	63	2D	1E	B8
CA	41	60	49	96	3C	5D	28	FB	BF	04	62	EB	AD	7F	9F
EE	AD	F0	DF	D4	75	3D	AC	91	AC	8E	F8	AC	1C	4B	4D
3C	78	F8	A5	43	58	52	0B	C6	A7	D0	44	94	CE	2C	4E
1F	2E	78	40	60	6A	DB	F9	27	5E	F7	61	FF	C1	3F	CC
A6	26	9B	0E	17	B2	E9	A4	4A	2D	84	59	68	7F	F3	D6
DC	C1	98	6D	33	E2	A1	12	16	20	95	DB	8F	A8	44	41
E3	**76**	17	C9	58	ED	6E	AC	68	4B	2F	09	6C	62	03	00
76	C9	3F	14	42	5E	73	73	B4	EA	8D	FF	46	CA	08	

We'll corrupt the message with three errors, 61h, 86h and 18h, in the highlighted positions of Table 8.7 which shows the message as it is seen by the receiver.

Table 8.8 Decoded frequency domain message after errors

2	(	i	à	°	H	f	¦	r	_	+	+	+	7	]	U
½	1	M	8	+	!	°	é	â	-	-	u	J	û	æ	_
D	[	D	G	?	¦	ò	3	(	§	¦	¦	ˋ	_	é	q
_	+	q	ò			Ñ	+	'	±	ù		É	+	¦	¢
A	<	9	ù	2	A	1	+	,	_	P	¦	9	¦	e	]
û	É			_	L	S		¦	¿	ê	5	n		'		m
	q	å	Ï	j	™	]		à	‡	;	ê	Ù			
Ä	©	Ò	A	8	u	‡	;		!	Ô	*	ó	f	Æ	#
×	&	J	>	†	3		S	S	N	{	ê	Ç		v	Y
Ô	¶	v	¢	‡	"	µ	Ð	d	\\	A		ð	Ë	)	ë
ó		'	—	®	Ó	R	}	0		2		)	ß	0	ô
Ð		9	™	e	Q			ö		T	S	F	3	ä	†
v	f		Ý	¾	g	ç	Æ	<	à	î	q	Ý	J		¥
Ê	)	û	ß	R	—	Ò	®	ì	R	8	;	Â	‰		<
Ã	!	ô	•	^	>	Ð	h	Y	Ó	—	û			S	N
z	©	,	8	q		'	¶	w	¢	,	9	ð	Ô	w	

Translation of the resulting message codes will depend on the fonts that are in use, but one thing should be clear from Table 8.8, and that is that the message is nothing like the original. At first sight it might appear that our attempt at error coding has left us worse off than if none had been used at all, but the important thing to remember is that we can get back to the correct message. The first 16 symbols should have been zero but they are in fact '2(ià°Hf|r_+++7]U' which in hexadecimal are

FD, 28, 69, 85, A7, 48, 9F, BA, 72, F5, B7, DA, BB, 37, 5D, 55

or

$$\alpha^{80}, \alpha^{53}, \alpha^{58}, \alpha^{128}, \alpha^{205}, \alpha^{226} \text{ etc.}$$

From these, we must reconstruct the entire error spectrum by using recursive extension. We have built enough protection into the message to correct up to eight erroneous symbols, but in this example only three were introduced. Since we know that this is the case we can cheat a little and solve for only three errors. The key equations give us the following

$$E_0.L_3 + E_1.L_2 + E_2.L_1 = E_3$$
$$E_1.L_3 + E_2.L_2 + E_3.L_1 = E_4$$
$$E_2.L_3 + E_3.L_2 + E_4.L_1 = E_5$$

Rearranging these into a matrix form gives

$$\begin{bmatrix} E_0 & E_1 & E_2 \\ E_1 & E_2 & E_3 \\ E_2 & E_3 & E_4 \end{bmatrix} \times \begin{bmatrix} L_3 \\ L_2 \\ L_1 \end{bmatrix} = \begin{bmatrix} E_3 \\ E_4 \\ E_5 \end{bmatrix}$$

Solving this by matrix inversion is actually quite economic since many of the terms are shared and no account need be taken of subtraction. The inverse matrix has the following form

$$\frac{1}{\Delta}\begin{bmatrix} E_2E_4 + E_3^2 & E_1E_4 + E_2E_3 & E_1E_3 + E_2^2 \\ E_1E_4 + E_2E_3 & E_0E_4 + E_2^2 & E_0E_3 + E_1E_2 \\ E_1E_3 + E_2^2 & E_0E_3 + E_1E_2 & E_0E_2 + E_1^2 \end{bmatrix} \times \begin{bmatrix} E_3 \\ E_4 \\ E_5 \end{bmatrix} = \begin{bmatrix} L_3 \\ L_2 \\ L_1 \end{bmatrix}$$

where

$$\Delta = E_0\left(E_2E_4 + E_3^2\right) + E_1\left(E_1E_4 + E_2E_3\right) + E_2\left(E_1E_3 + E_2^2\right)$$

Using the solution from this to construct a recursive extension circuit with three elements, we can complete the error spectrum which appears as follows in Table 8.9.

The received (corrupted) message, after conversion back into the frequency domain (shown previously in Table 8.8), shares the top line (albeit as ASCII symbols) of Table 8.9. This tells us that our assumption that there were only three errors and subsequent use of only the first six elements to form the key equations was correct. Had there been more or less errors only the first six recursively extended elements would have agreed with the top line of Table 8.8. Converting this into the time domain by inverse Fourier transforming gives the original errors that corrupted the time domain message. These are given in Table 8.10 with all non-zero values shown white on black.

Table 8.9 Completed error spectrum using recursive extension

FD	28	69	85	A7	48	9F	BA	72	F5	B7	DA	BB	37	5D	55
E2	57	6D	4C	B1	44	87	F0	E6	4A	B9	06	3E	F3	E3	93
64	3A	36	22	1F	C4	E7	56	05	79	B2	D7	04	82	E6	51
95	BD	05	FD	28	69	85	A7	48	9F	BA	72	F5	B7	DA	BB
37	5D	55	E2	57	6D	4C	B1	44	87	F0	E6	4A	B9	06	3E
F3	E3	93	64	3A	36	22	1F	C4	E7	56	05	79	B2	D7	04
82	E6	51	95	BD	05	FD	28	69	85	A7	48	9F	BA	72	F5
B7	DA	BB	37	5D	55	E2	57	6D	4C	B1	44	87	F0	E6	4A
B9	06	3E	F3	E3	93	64	3A	36	22	1F	C4	E7	56	05	79
B2	D7	04	82	E6	51	95	BD	05	FD	28	69	85	A7	48	9F
BA	72	F5	B7	DA	BB	37	5D	55	E2	57	6D	4C	B1	44	87
F0	E6	4A	B9	06	3E	F3	E3	93	64	3A	36	22	1F	C4	E7
56	05	79	B2	D7	04	82	E6	51	95	BD	05	FD	28	69	85
A7	48	9F	BA	72	F5	B7	DA	BB	37	5D	55	E2	57	6D	4C
B1	44	87	F0	E6	4A	B9	06	3E	F3	E3	93	64	3A	36	22
1F	C4	E7	56	05	79	B2	D7	04	82	E6	51	95	BD	05	

Table 8.10 Reconstructed errors extracted from the error spectrum

00	00	00	00	00	00	00	00	00	00	00	00	00	00	00	00
00	00	00	00	00	00	00	00	00	00	00	00	00	00	00	00
00	00	00	00	00	00	00	00	00	00	00	00	00	00	00	00
00	00	00	00	00	00	00	00	00	00	00	00	00	00	00	00
00	00	00	00	00	00	61	00	00	00	00	00	00	00	00	00
00	00	00	00	00	00	00	00	00	00	00	00	00	00	00	00
00	00	00	00	00	00	00	00	00	00	00	00	00	00	00	00
00	00	00	00	00	00	00	00	84	00	00	00	00	00	00	00
00	00	00	00	00	00	00	00	00	00	00	00	00	00	00	00
00	00	00	00	00	00	00	00	00	00	00	00	00	00	00	00
00	00	00	00	00	00	00	00	00	00	00	00	00	00	00	00
00	00	00	00	00	00	00	00	00	00	00	00	00	00	00	00
00	00	00	00	00	00	00	00	00	00	00	00	00	00	00	00
00	00	00	00	00	00	00	00	00	00	00	00	00	00	00	00
00	18	00	00	00	00	00	00	00	00	00	00	00	00	00	00
00	00	00	00	00	00	00	00	00	00	00	00	00	00	00	

A quick comparison with the table above, and Tables 8.6 and 8.7 will show that the errors do represent the difference between the transmitted and received messages. The Pascal program used to carry out this experiment is listed in Appendix C. Also included in Appendix C are two further examples. The second is a program for correcting up to eight errors in the above message making use of an automatic rather than long-hand matrix inversion for the solution of the key equations. The third example again corrects eight errors but uses a matrix inversion optimized for the key equations. This takes somewhat under one-tenth of the time of the generalized inversion.

Using fractionally over **half** the redundancy required by simple parity we have generated a correction scheme capable of correcting up to 64 bits spread over eight symbols.

8.8 EXCEEDING THE ERROR CAPACITY OF THE CODE

The examples presented so far have been done with the benefit of a little foreknowledge. Apart from knowing what the errors were, the solutions were performed with a knowledge of how many errors were present. In reality, of course, this information is not usually available unless it can be gleaned from careful packaging of the message (see later). If a correction is attempted for the wrong number of errors then the result will be incorrect. If t errors occur then the key equations will be solved with t linearly independent equations. An attempt to perform a matrix inversion on a set of equations that are not all linearly independent will fail, so we are alerted to our mistake.

What is less obvious is what happens when more errors occur than our code allows for. There is always the chance that the errors will combine to make the code appear to produce a correct solution for a different message. With this kind of coding, however, there are no-man's lands between valid codes which appear no more or less like another valid message than the message from which they originated. Technically the reason is because this is not a **perfect** code. The Hamming code discussed earlier is an example of a perfect code because all possible messages will decode to a solution. If the code's error capacity is exceed, we can't know it. The use of expurgation (see error correction using the CRC) to increase d_{min} from three to four is a mechanism for creating these undefined states. Where two bit errors occur, while we don't know what the correct message was we are aware that our ability to correct it has been exceeded.

If we introduce more errors than our coding allows and solve the key equations for the maximum number of errors, inevitably the $2t$ known error spectra re-emerge from the recursive extension circuit, but upon inverse Fourier transforming the completed spectrum we don't end up with t errors (non-zero results). If we add a fourth error (35H at position A0H) to the received message shown in Table 8.7, then the decoded error spectrum appears as Table 8.11. Since our solution allowed only three errors to be present, the 253 errors that appear alert us to a problem. It is, however, possible to detect this case before wasting time inverse Fourier transforming the spectrum.

Table 8.11 The reconstructed error vector after exceeding the code's capacity

64	8A	FA	24	90	50	8A	BD	C4	A1	97	08	38	04	BC	7C
AD	50	70	B3	EA	E1	9D	55	97	31	2A	63	5D	65	60	BD
E5	6B	BC	FF	05	B2	AC	C4	18	54	CA	A8	07	0D	36	F3
03	09	F9	B8	A4	67	7A	D9	DE	5A	A9	6A	E5	31	AF	5E
B7	1A	BF	09	4A	01	44	FC	98	65	AD	E7	A8	A0	A3	00
D5	5E	FA	B4	77	86	92	7A	EA	EF	88	EC	E1	5D	0E	A1
49	9A	44	20	9F	5A	B2	80	BB	01	DC	DB	1C	AC	CA	FD
4C	25	DE	10	49	4E	F0	C7	DC	FF	86	20	D2	25	3B	8B
03	D2	BE	A4	BF	83	86	71	B6	87	3F	05	AF	DB	C1	BE
B4	12	C7	9F	B3	90	13	7C	04	C5	1E	F5	6A	56	43	92
8F	38	12	4C	D1	F5	24	18	8F	79	71	EF	DD	D1	98	13
E7	1C	D5	33	C1	73	9A	36	A0	48	2F	4F	BB	85	82	9B
80	63	DD	43	48	2A	33	87	85	67	77	6B	1E	F4	4F	F9
9D	D9	7B	A9	02	9B	54	6D	C5	C0	88	8B	06	02	70	57
F0	C0	07	82	83	FC	06	73	79	4E	10	7B	3B	FD	60	08
57	F3	64	F4	55	4A	0E	B6	2F	3F	6D	0D	56	B7	A3	

A valid error spectrum will be cyclic over the period of the message. Our code allows for eight errors although we only attempted correction over three. The 16 known error spectra for the four-error case are

C8, 26, 64, BF, 42, C1, 64, F8, 9C, 28, 0A, C9, 4A, C7, 4B, FD

The first 16 codes generated by the recursive extension circuit are as follows:

C8, 26, 64, BF, 42, C1, 8F, 3A, 1F, 78, 04, 0E, 10, 59, 6E, EB

Only the first six of 16 elements (shown underlined) agree with the known error spectrum. These six elements are of course those used in the construction of the key equations for solving three errors. The ten remaining known error spectra tell us of our mistake in solving for only three errors. What if nine symbol errors had occurred and we had used all 16 elements in the solution, solving for all eight correctable errors? In this case there are no known error spectra remaining with which to make such a comparison. If we use all 16 elements in the solution of the key equations we guarantee that at least those 16 will be reproduced by the recursive extension circuit. It would seem at first that the only way to see if more than eight errors have occurred is to perform the IFT and see if there are more than t errors in the reconstructed error vector as in Table 8.11. The cyclic nature of the REC, however, allows us to avoid this.

In the example here, the message is 255 symbols long. If the REC is clocked 255 times then for a valid error spectrum, the contents of its registers will be returned to their starting state – the first t error spectra. If the error spectrum has more roots than the REC can generate then its register contents will be changed. Continuing with this four-error example, shown below are the first three elements produced by the REC the first time around, and below them, their values 255 clocks later. If there had been only three errors present (matching our attempted solution), both vectors would be identical. So without resorting to an extra IFT, we can know that our attempted solution is incorrect.

Initial register contents

C8, 26, 64

Register contents after 255 clocks

70, 9F, 95

This is a bit like trying to generate a cubic curve from a quadratic equation.

8.9 DISCUSSION

Use of the frequency domain opens up the possibility of applying regular transform-like operations on our data for the purposes of performing error protection and correction. It should be apparent, however, that this is not without cost. While the processing is regular and can be easily extended to different levels of protection within the same hardware, there is a deal more of it. Even where no errors have occurred a Fourier transform is required in order to recover the original data from the transmission. If errors have occurred then a further two such transforms will be needed as well as a possible matrix inversion and recursive extension.

Where a microprocessor is available many of these complexities can be pushed into software, albeit at the expense of speed, but it is these issues which mean that the story is by no means over. While we have covered the basic mathematical tools which are available to perform error coding, by cunning and appropriate application, much better use can be made of them than the examples in this chapter have done. To this end, the next few chapters deal with strategies that allow us to get the most from our coding.

9

Mixed domain error coding

9.1 ADVANTAGES OF TIME DOMAIN PROCESSING

You will probably have noticed that in many ways time and frequency domain treatment of error correction is very similar. If we Fourier transform a suitably encoded time domain message the transformed vector will contain $2t$ consecutive zeros. If errors occur in a time domain encoded message the problem that requires solving is identical to that which must be solved in the frequency domain case, only it is treated very differently.

There are two key advantages of time domain coding that may make it attractive in some circumstances. First, time domain treatment of messages requires less processing in the event of no errors than does frequency domain. With frequency domain coding a minimum of one Fourier transform must be computed before the original data are recovered, whereas with time domain coding the data are present as they arrive. Also, with time domain coding the presence or absence of errors is known almost instantly after the last symbol of the block is received. A second major advantage of time domain processing is that in the event that the correctional capacity of the code is exceeded, all may not be lost. If the data are textual or perhaps speech or images, errors may be recognized by other means or even ignored, and the message might still be useful. With frequency domain coding if the capacity of the code is exceeded there is no simple route back to even an approximation of the original data.

9.2 CALCULATING THE ERRORS IN THE FREQUENCY DOMAIN

If time domain processing is to be used, calculation of the $2t$ check symbols is unavoidable, but the excruciating maths involved in correcting a corrupted time domain message can be handled very differently. If we consider the double-correcting example of Chapter 7, the time domain vector

$$3, 5, 7, \mathbf{2}, \mathbf{5}, \mathbf{5}, \mathbf{3}$$

was produced where the last four symbols (bold) are check symbols. Performing a Fourier transform on this yields the vector

$$0, 4, 3, 4, 0, 0, 0$$

with four zeros. Suppose that the time domain vector suffers the two errors shown below

$$0, 0, 6, 0, 0, 3, 0$$

so that it appears as

$$3, 5, \mathit{1}, 2, 5, \mathit{6}, 3$$

at the receiver. The Fourier transform of the received and corrupted message is now

$$\mathbf{5}, 3, 4, 1, \mathbf{3}, \mathbf{0}, \mathbf{3}$$

where the bold numbers should have been zero. Using these known error spectra we can construct an REC. If for convenience we rearrange the vector by symbol rotating as

$$3, 0, 3, 5, 3, 4, 1$$

and perform recursive extension using the first four elements, we get the complete (but rotated) error vector

$$3, 0, 3, 5, 7, 7, 5$$

Rearranging the symbols back to their original positions gives

$$5, 7, 7, 5, 3, 0, 3$$

and the inverse Fourier transform of this is

$$0, 0, 6, 0, 0, 3, 0$$

the error vector. Last, the error vector is XORed with the received message to correct it.

Symbol rotating is not so much a necessity as a convenience. The only requirement of constructing the REC is that the known error spectra should be adjacent in the vector. In the example of Chapter 7 the known error spectra cross the boundary between the start and end of the message such that one of them appears to be separated from the other three. The vector produced by the REC repeats continuously after each seven symbols so the last symbol is actually adjacent to the first. The act of symbol rotating simply serves to make this problem look like the frequency domain examples of Chapter 8. If the message is constructed with this solution in mind, then rotating will not be necessary.

To check for errors we still need to calculate the syndromes as the message arrives, and it transpires that if the message is constructed correctly, these actually provide the starting conditions of the REC, or the error spectra that should have been zero. This means that only an inverse Fourier transform is needed to translate the completed error spectrum back into the time domain after recursive extension.

Using this hybrid approach we need only to call on time-costly processing in the event of errors, so speeding up decoding. Even in the event of errors decoding will be faster by two Fourier transforms than a full frequency domain approach. We also have a fallback position if the capacity of the code is exceeded which may be essential if the data cannot be retransmitted. We still have to consider the original encoding of the message if this approach is to take on a protection independent form, however.

9.3 FINDING THE CHECK SYMBOLS USING MATRIX INVERSION

In Chapters 6 and 7, calculation of the time domain check symbols was done by solving simultaneous equations involving all data symbols and all check symbols. This much of the problem can be simplified by using a process a little like erasure, mentioned in Section 7.1. The check symbols are set to zero and the syndromes are calculated exactly as they would be upon decoding at the receiver. For a double-symbol-correcting time domain message over GF(2^3), the message should satisfy the four conditions imposed by (9.1):

$$\sum_{i=0}^{6} d_i \alpha^{ji} = 0 \quad (0 \le j \le 3) \tag{9.1}$$

where d_{0-3} are the as yet unknown check symbols while d_{4-6} are the data. Since the check symbols are initially zero we actually end up with four syndromes S_{0-3} from (9.2).

$$\sum_{i=0}^{6} d_i \alpha^{ji} = S_j \quad (0 \le j \le 3) \tag{9.2}$$

From (9.2) S_{0-3} have the form shown below:

$$S_0 = d_0 + d_1 + d_2 + d_3$$
$$S_1 = d_0 + \alpha d_1 + \alpha^2 d_2 + \alpha^3 d_3$$
$$S_2 = d_0 + \alpha^2 d_1 + \alpha^4 d_2 + \alpha^6 d_3$$
$$S_3 = d_0 + \alpha^3 d_1 + \alpha^6 d_2 + \alpha^2 d_3$$

Which can be converted into the matrix format below

$$\begin{bmatrix} 1 & 1 & 1 & 1 \\ 1 & \alpha & \alpha^2 & \alpha^3 \\ 1 & \alpha^2 & \alpha^4 & \alpha^6 \\ 1 & \alpha^3 & \alpha^6 & \alpha^2 \end{bmatrix} \begin{bmatrix} d_0 \\ d_1 \\ d_2 \\ d_3 \end{bmatrix} = \begin{bmatrix} S_0 \\ S_1 \\ S_2 \\ S_3 \end{bmatrix}$$

From this final form the time domain check symbols d_{0-3} are easily found by inverting the matrix, and without the need to involve all the data symbols. Over GF(2^3) there are only three data symbols in any case, but for realistic blocks of, say, 255 symbols, this approach makes the maths far easier. The above solution will be true for a double-correcting code with check symbols in data positions 0 to 3, regardless of the block size, except that α^9 in this field (bottom right-hand corner of the matrix) has folded over to become α^2. We are thus approaching some sort of regular format to time domain coding. Such a generalized form is shown below, where t symbol errors are to be protected against.

$$\begin{bmatrix} \alpha^0 & \alpha^0 & \alpha^0 & \cdots & \alpha^0 \\ \alpha^0 & \alpha^1 & \alpha^2 & \cdots & \alpha^{2t-1} \\ \alpha^0 & \alpha^2 & \alpha^4 & \cdots & \alpha^{4t-2} \\ \vdots & \vdots & \vdots & \ddots & \vdots \\ \alpha^0 & \alpha^{2t-1} & \alpha^{4t-2} & \cdots & \alpha^{(2t-1)^2} \end{bmatrix} \begin{bmatrix} d_0 \\ d_1 \\ d_2 \\ \vdots \\ d_{2t-1} \end{bmatrix} = \begin{bmatrix} S_0 \\ S_1 \\ S_2 \\ \vdots \\ S_{2t-1} \end{bmatrix}$$

or if the element positions are given by (i, j) then we have

$$\left[\alpha^{ij}\right]\left[d_j\right] = \left[S_j\right] \quad (0 \le i, j \le 2t-1)$$

The solution to the previous example for $t = 2$ over $GF(2^3)$ is

$$\begin{bmatrix} \alpha^3 & \alpha^5 & \alpha^3 & \alpha^4 \\ \alpha^5 & 0 & \alpha^4 & \alpha^7 \\ \alpha^3 & \alpha^4 & 0 & \alpha^6 \\ \alpha^4 & \alpha^7 & \alpha^6 & \alpha \end{bmatrix} \begin{bmatrix} S_0 \\ S_1 \\ S_2 \\ S_3 \end{bmatrix} = \begin{bmatrix} d_0 \\ d_1 \\ d_2 \\ d_3 \end{bmatrix}$$

Notice that the forward and inverted matrices are symmetrical which means that we need solve only half the inversion. Since the left-most matrix does not depend on the data the results may be precalculated and maintained in tables, accessed when the level of protection is programmed. To illustrate the complete process, using the data 3, 6, 1 (d_{4-6}), we construct an initial message vector

$$0, 0, 0, 0, 3, 6, 1$$

From (9.2) four syndromes are calculated, which in this example are

$$\alpha^2, 0, 0, \alpha^3 \ (S_{0-3})$$

Multiplying these by the matrix above gives

$$\alpha^4, \alpha^1, \alpha^3, \alpha^3 \ (d_{0-3})$$

The completed message is, therefore,

$$6, 2, 3, 3, 3, 6, 1$$

We can verify this by recalculating the syndromes from (9.2). If the check symbols are correct then all the syndromes will be zero. You can try this for yourself. Appendix F lists a more extensive example using this technique where four symbol errors may be located and corrected in a 255-symbol message over $GF(2^8)$. To summarize this process, the following steps are required:

- Set all check symbols in the message to zero
- Calculate the syndromes just as the decoder or receiver would
- Multiply the syndromes by the (precalculated) solution matrix
- Replace the check symbol zeros with the results of the multiply.

9.4 FINDING THE CHECK SYMBOLS USING THE FFT

It is possible to calculate the check symbols using the decoder if a Fourier transformer is available. In some cases this may reduce the amount of hardware that is required and this is important on low-cost chip designs. By setting the check symbols to zero we effectively have a message with four unknown errors in four known positions. If $t = 2$, for example, we can solve for four unknowns. Normally these unknowns will be two error positions and their corresponding patterns, but we can trade the two positions for two further patterns since we know where the errors are, i.e. in the positions of the check symbols.

In Section 8.3 it was demonstrated that for an REC capable of completing the error spectrum of two errors, the feedback multipliers, L, were defined by the quadratic $\alpha^{2x} + \alpha^x L_1 + L_2 = 0$ where x has two solutions corresponding to the positions of the two errors. If the two positions are known then L_1 and L_2 can be found directly, requiring the knowledge of only two error spectra in order to complete the spectrum. In our case where $t = 2$ we have four errors (equal to the check symbols) so we must construct an REC of the form

$$\alpha^{4x} + L_1\alpha^{3x} + L_2\alpha^{2x} + L_3\alpha^x + L_4 = 0$$

Knowing the values of x (0 to 3), from

$$\left(\alpha^x + \alpha^0\right)\left(\alpha^x + \alpha^1\right)\left(\alpha^x + \alpha^2\right)\left(\alpha^x + \alpha^3\right) = 0$$

we know that

$$L_1 = \alpha^0 + \alpha^1 + \alpha^2 + \alpha^3$$

$$L_2 = \alpha^{0+1} + \alpha^{0+2} + \alpha^{0+3} + \alpha^{1+2} + \alpha^{1+3} + \alpha^{2+3}$$
$$= \alpha^1 + \alpha^2 + \alpha^4 + \alpha^5$$

$$L_3 = \alpha^{0+1+2} + \alpha^{0+1+3} + \alpha^{0+2+3} + \alpha^{1+2+3}$$
$$= \alpha^3 + \alpha^4 + \alpha^5 + \alpha^6$$

$$L_4 = \alpha^{0+1+2+3}$$
$$= \alpha^6$$

These are easy to compute and allow us to fix up the REC. Using the primitive polynomial $x^3 + x + 1 = 0$, L_{1-4} are α^2, α^5, α^5, α^6. Next, the REC must be preloaded. Since we are solving for four errors, the REC will have four registers. To find the values with which to preload the REC a Fourier transform must be performed on the message.

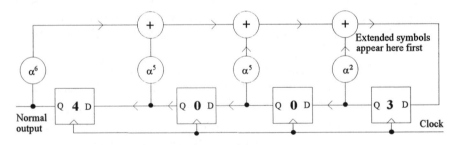

Fig. 9.1 Using the REC to calculate the check symbols.

Using the previous example where the initial message is

$$0,\, 0,\, 0,\, 0,\, 3,\, 6,\, 1$$

then the Fourier transform of this is

$$4, 0, 0, 3, 7, 2, 2$$

Preloading the REC with 4, 0, 0, 3 (the four syndromes in the previous section), we have the following circuit, given in Fig. 9.1. Clocking the circuit two times yields the three unknown check symbol spectra at the input to the right-hand register as follows:

$$\alpha^6\alpha^2 + \alpha^5 0 + \alpha^5 0 + \alpha^2\alpha^3 = \alpha^6 \quad (5)$$

As the circuit appears in Fig. 9.1,

$$\alpha^6 0 + \alpha^5 0 + \alpha^5\alpha^3 + \alpha^2\alpha^6 = 0 \quad (0)$$

After one clock and

$$\alpha^6 0 + \alpha^5\alpha^3 + \alpha^5\alpha^6 + \alpha^2 0 = \alpha^2 \quad (4)$$

after a second clock. The completed spectrum is, therefore,

$$4, 0, 0, 3, 5, 0, 4$$

Performing an inverse Fourier transform in this gives the vector

$$6, 2, 3, 3, 0, 0, 0$$

This represents the time domain errors which in this case are the wanted check symbols, so the completed message is found by XORing the error vector with the original vector giving 6, 2, 3, 3, 3, 6, 1. This is the same solution that we achieved in Section 9.3. You will notice that the initial Fourier transform need only compute $2t$ spectra (sufficient to preload the REC) which in some implementations will speed up the calculation. To summarize, the steps involved are as follows:

- Set the check symbols to zero
- Set up an REC based on the check symbol positions
- Fourier transform the message
- Preload the REC using the appropriate spectra

- Use the REC to complete the spectrum
- Inverse Fourier transform the completed check symbol spectrum
- Add the check symbols to the message and transmit.

9.5 DISCUSSION

In this chapter we have looked at some possible advantages of transmitting messages in the time domain. These include less processing, potential recovery in the event of too many errors and much faster decoding both with or without errors. The problem has been split into two parts, calculating the check symbols and locating and correcting errors. In the event of non-zero syndromes at the receiver decoding is identical to that performed in the frequency domain except that one less Fourier transform is required. Where no errors are detected the data are available with no further processing.

In order to generalize the problem, two suggestions are presented for the initial calculation of the check symbols at the encoder. These avoid the need to establish unique solutions dependent on the protection required. The first solution is fast making the check symbols available shortly after the arrival of the data in the encoder, while the second approach, though slower and involving the use of two Fourier transforms, makes extensive re-use of existing hardware necessary for frequency domain decoding.

As the designer, you are again faced with options and possibilities from which you can draw in order to meet your particular system requirements.

Part Five

Packaging the data

10

Block-interleaving

It is important to consider error correction in conjunction with the intended application since this has a strong bearing on the way in which the data should be packaged for best effect. There are situations where so much data can be lost at a time, that to provide a code capable of handling the volume of errors would be impractical. This does not, however, mean that the data need be lost if the error codes are correctly packaged. A particular example where gross loss of data can occur is magnetic storage on tape. Because very high-bit densities can be associated with tape, even a small failure of the medium will wipe out many consecutive symbols.

To offset this problem a process called **block-interleaving** can be used. There are many variations on this theme but all follow the same principle. The essence of this process is to reorder the sequencing of data as it is stored onto the recording medium (or transmission channel). The simplest form of reordering is to construct a two-dimensional block of data, much like the tables of previous examples, building up the block row by row. When the block is complete it is read out onto the medium or channel, column by column. As far as the example of Chapter 8 is concerned there would be no advantage in doing this so block-interleaving must be done in conjunction with a reorganization of the error control codes.

Suppose that each row of the data block was a self-contained group of error coded data capable of recovering, say, two errors. If we lose any more than two symbols per row then the data is irreparably damaged, but by writing the data to the channel in columns, large consecutive losses will take only small chunks out of each encoded data group. This is illustrated in Table 10.1.

The data are read into the block horizontally and in this example each row can correct one error. The data are written into the channel vertically and the shaded blocks represent 11 consecutively corrupted symbols. On reconstruction of the block at the receiver or on recovery from the storage medium, each row has at most one error in and can therefore be corrected.

In fact we could lose up to 16 symbols in a row and still recover from the situation if no further symbols are lost. The amount of redundancy present here is the same as if we were to correct 16 errors in the whole block regardless of position but the major difference is that the processing necessary to achieve this cover extends only to a single correcting code.

Table 10.1 Generating a block-interleaved message

Data														Check	
d_{00}	d_{01}	d_{02}	d_{03}	d_{04}	d_{05}	d_{06}	d_{07}	d_{08}	d_{09}	d_{0A}	d_{0B}	d_{0C}	p_{0D}	q_{0E}
d_{10}	d_{11}	d_{12}	d_{13}	d_{14}	d_{15}	d_{16}	d_{17}	d_{18}	d_{19}	d_{1A}	d_{1B}	d_{1C}	p_{1D}	q_{1E}
d_{20}	d_{21}	d_{22}	d_{23}	d_{24}	d_{25}	d_{26}	d_{27}	d_{28}	d_{29}	d_{2A}	d_{2B}	d_{2C}	p_{2D}	q_{2E}
d_{30}	d_{31}	d_{32}	d_{33}	d_{34}	d_{35}	d_{36}	d_{37}	d_{38}	d_{39}	d_{3A}	d_{3B}	d_{3C}	p_{3D}	q_{3E}
d_{40}	d_{41}	d_{42}	d_{43}	d_{44}	d_{45}	d_{46}	d_{47}	d_{48}	d_{49}	d_{4A}	d_{4B}	d_{4C}	p_{4D}	q_{4E}
d_{50}	d_{51}	d_{52}	d_{53}	d_{54}	d_{55}	d_{56}	d_{57}	d_{58}	d_{59}	d_{5A}	d_{5B}	d_{5C}	p_{5D}	q_{5E}
d_{60}	d_{61}	d_{62}	d_{63}	d_{64}	d_{65}	d_{66}	d_{67}	d_{68}	d_{69}	d_{6A}	d_{6B}	d_{6C}	p_{6D}	q_{6E}
d_{70}	d_{71}	d_{72}	d_{73}	d_{74}	d_{75}	d_{76}	d_{77}	d_{78}	d_{79}	d_{7A}	d_{7B}	d_{7C}	p_{7D}	q_{7E}
d_{80}	d_{81}	d_{82}	d_{83}	d_{84}	d_{85}	d_{86}	d_{87}	d_{88}	d_{89}	d_{8A}	d_{8B}	d_{8C}	p_{8D}	q_{8E}
d_{90}	d_{91}	d_{92}	d_{93}	d_{94}	d_{95}	d_{96}	d_{97}	d_{98}	d_{99}	d_{9A}	d_{9B}	d_{9C}	p_{9D}	q_{9E}
d_{A0}	d_{A1}	d_{A2}	d_{A3}	d_{A4}	d_{A5}	d_{A6}	d_{A7}	d_{A8}	d_{A9}	d_{AA}	d_{AB}	d_{AC}	p_{AD}	q_{AE}
d_{B0}	d_{B1}	d_{B2}	d_{B3}	d_{B4}	d_{B5}	d_{B6}	d_{B7}	d_{B8}	d_{B9}	d_{BA}	d_{BB}	d_{BC}	p_{BD}	q_{BE}
d_{C0}	d_{C1}	d_{C2}	d_{C3}	d_{C4}	d_{C5}	d_{C6}	d_{C7}	d_{C8}	d_{C9}	d_{CA}	d_{CB}	d_{CC}	p_{CD}	q_{CE}
d_{D0}	d_{D1}	d_{D2}	d_{D3}	d_{D4}	d_{D5}	d_{D6}	d_{D7}	d_{D8}	d_{D9}	d_{DA}	d_{DB}	d_{DC}	p_{DD}	q_{DE}
d_{E0}	d_{E1}	d_{E2}	d_{E3}	d_{E4}	d_{E5}	d_{E6}	d_{E7}	d_{E8}	d_{E9}	d_{EA}	d_{EB}	d_{EC}	p_{ED}	q_{EE}

Some systems augment this interleaving by adding a set of **inner** Reed–Solomon codes to the block. The **outer** codes refer to those initially embedded into the rows. If we use time domain coding vertically, then the bottom two rows of encoded data could be replaced by Reed–Solomon check symbols. We now have a vertical and horizontal correction capability. To enhance this capability, the data can be written into the media diagonally. In this way, even in the presence of large errors it is often the case that the capacities of neither horizontal nor vertical codes to correct the block are exceeded.

Again, in Table 10.2 the shaded boxes represent consecutive errors but no more than one error appears in either a row or a column. The inner and outer codes augment each other in a rather useful way. Large burst errors will render the inner code ineffective, but will be so distributed as to allow the outer codes to still function. The inner codes, therefore, deal with small random errors. Considering Table 10.1, while there is only one error per row due to the burst error, a further random error in the wrong place may well destroy the chances of complete recovery. The inner codes reduce this risk. So the outer codes deal with large losses of data, decorrelated by block-interleaving, while the inner codes mop up small random errors which could otherwise compromise our ability to recover from a large error.

Table 10.2 Diagonal block-interleaving with inner and outer codes

Data													Check	
d_{00}	d_{01}	d_{02}	d_{03}	d_{04}	d_{05}	d_{06}	d_{07}	d_{08}	d_{09}	d_{0A}	d_{0B}	d_{0C}	P_{0D}	q_{0E}
d_{10}	d_{11}	d_{12}	d_{13}	d_{14}	d_{15}	d_{16}	d_{17}	d_{18}	d_{19}	d_{1A}	d_{1B}	d_{1C}	P_{1D}	q_{1E}
d_{20}	d_{21}	d_{22}	d_{23}	d_{24}	d_{25}	d_{26}	d_{27}	d_{28}	d_{29}	d_{2A}	d_{2B}	d_{2C}	P_{2D}	q_{2E}
d_{30}	d_{31}	d_{32}	d_{33}	d_{34}	d_{35}	d_{36}	d_{37}	d_{38}	d_{39}	d_{3A}	d_{3B}	d_{3C}	P_{3D}	q_{3E}
d_{40}	d_{41}	d_{42}	d_{43}	d_{44}	d_{45}	d_{46}	d_{47}	d_{48}	d_{49}	d_{4A}	d_{4B}	d_{4C}	P_{4D}	q_{4E}
d_{50}	d_{51}	d_{52}	d_{53}	d_{54}	d_{55}	d_{56}	d_{57}	d_{58}	d_{59}	d_{5A}	d_{5B}	d_{5C}	P_{5D}	q_{5E}
d_{60}	d_{61}	d_{62}	d_{63}	d_{64}	d_{65}	d_{66}	d_{67}	d_{68}	d_{69}	d_{6A}	d_{6B}	d_{6C}	P_{6D}	q_{6E}
d_{70}	d_{71}	d_{72}	d_{73}	d_{74}	d_{75}	d_{76}	d_{77}	d_{78}	d_{79}	d_{7A}	d_{7B}	d_{7C}	P_{7D}	q_{7E}
d_{80}	d_{81}	d_{82}	d_{83}	d_{84}	d_{85}	d_{86}	d_{87}	d_{88}	d_{89}	d_{8A}	d_{8B}	d_{8C}	P_{8D}	q_{8E}
d_{90}	d_{91}	d_{92}	d_{93}	d_{94}	d_{95}	d_{96}	d_{97}	d_{98}	d_{99}	d_{9A}	d_{9B}	d_{9C}	P_{9D}	q_{9E}
d_{A0}	d_{A1}	d_{A2}	d_{A3}	d_{A4}	d_{A5}	d_{A6}	d_{A7}	d_{A8}	d_{A9}	d_{AA}	d_{AB}	d_{AC}	P_{AD}	q_{AE}
d_{B0}	d_{B1}	d_{B2}	d_{B3}	d_{B4}	d_{B5}	d_{B6}	d_{B7}	d_{B8}	d_{B9}	d_{BA}	d_{BB}	d_{BC}	P_{BD}	q_{BE}
d_{C0}	d_{C1}	d_{C2}	d_{C3}	d_{C4}	d_{C5}	d_{C6}	d_{C7}	d_{C8}	d_{C9}	d_{CA}	d_{CB}	d_{CC}	P_{CD}	q_{CE}
r_{D0}	r_{D1}	r_{D2}	r_{D3}	r_{D4}	r_{D5}	r_{D6}	r_{D7}	r_{D8}	r_{D9}	r_{DA}	r_{DB}	r_{DC}	r_{DD}	r_{DE}
s_{E0}	s_{E1}	s_{E2}	s_{E3}	s_{E4}	s_{E5}	s_{E6}	s_{E7}	s_{E8}	s_{E9}	s_{EA}	s_{EB}	s_{EC}	s_{ED}	s_{EE}

By noting where errors occur in both columns and rows we can to an extent locate errors before attempting to correct them. Referring back to previous examples, if we know where the errors are, then we can correct twice as many as the code would initially suggest. This interaction between the inner and outer codes gives rise to synergy, where the combined actions of the codes is greater than we would expect from the addition of their actions apart.

Interleaving can be done in a number of ways, the most obvious being **bit-wise** and **symbol-wise**. With bit-wise interleaving the symbols are arranged horizontally, and during readout the columns are selected on a bit basis. An eight-bit symbol is spread over eight columns. With symbol-wise interleaving the symbols are actually vertical, so one column is one symbol wide. Normally, symbol-wise interleaving will be used since it usually gives rise to a more economic solution. Bit-wise interleaving does, however, have an interesting property.

Where enough consecutive errors occur to cause two adjacent bits to be corrupted on any single row, those two bits could cross a symbol boundary, giving rise to a two-symbol error on that row. If each symbol is m bits, then a further $m - 1$ bit columns can be corrupted before we risk creating a three symbol error. In other words, a two-symbol per row correction capability, when used with bit-wise interleaving, can correct up to $m + 1$ adjacent columns in error, compared with only two columns when symbol-wise interleaving is used. Of course you never get something for nothing, and you have to pay for this in terms of redundancy and processing. What the bit-wise interleaving does, however, is make a two-symbol correction process go much further.

A last consideration is the type of error coding that should be used with block-interleaving. There is no real constraint on the outer codes. These may be encoded in either the frequency or the time domain. The inner codes, however, should be encoded using only the time domain. The reasons for this are fairly obvious but stem from the fact that the large errors against which the outer codes protect will almost certainly exceed the capacity of the inner codes to correct. If frequency domain coding is used for the inner codes, and the capacity of the code is exceeded, then the whole column is lost since the original message cannot even be recovered in part. While the outer codes can still recover from this, it compromises their performance in the event of further errors elsewhere in the block. If time domain coding is used, then **only** the rows where column bits are in error will contain errors. Other rows will still contain no errors.

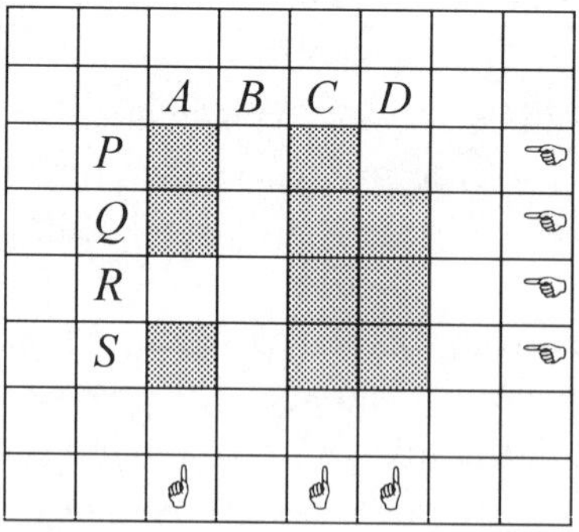

Fig. 10.1 Recovering from excess errors.

Figure 10.1 illustrates part of a block of data protected by single symbol correcting inner and outer codes. The shaded boxes represent a symbol error while the fingers point to rows and columns containing errors. In all cases the correcting capacities of the affected columns and rows have been exceeded. If we know where the errors are, however, then we can correct two errors per row or column using the single-symbol code. Non-zero syndromes in the outer codes tell us that errors have occurred in the rows labelled *P–S* while non-zero inner code syndromes indicate that the errors are in columns *A*, *C* and *D*.

Starting with row *P*, we know that the row contains more than one error, otherwise the outer code would correct it directly. We also know that it contains three or less errors since there are only three columns containing errors. This means that there are three possible combinations of two errors and one possible combination of three errors. By trial and error we can test the three possible error pairs *AC*, *AD* and *BC* to see if any result in zero syndromes. In this case *AC* will work out since there are only two errors in this row. Repeating for the remaining three rows in error, we will also be successful with row *R* which also contains only two errors.

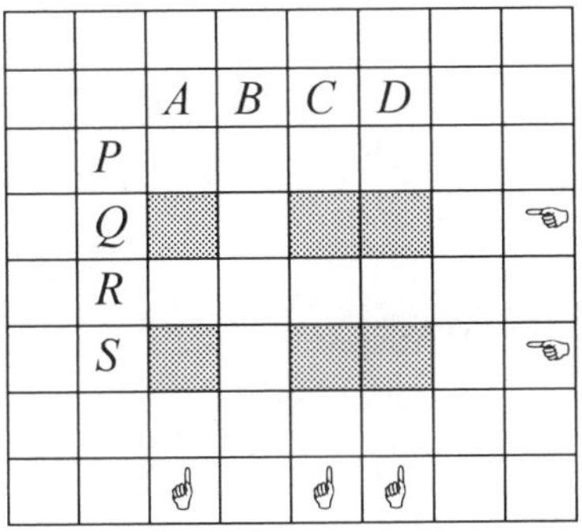

Fig. 10.2 Errors after horizontal correction.

Figure 10.2 shows the situation after horizontal correction. We can check each column to see if any exhibit only a single error now and, if not, we know that all the errors are two-symbol, and we know the symbol positions so we can correct the message. It may be that some results are indeterminate due to the extent of the errors, but we may still be able to find the correct solution by comparing the possible horizontal candidates with the possible vertical candidates. This is a contrived example, but it serves to illustrate the synergy between the inner and outer codes exceeding our expectations of the code capacity.

10.1 DISCUSSION

By using block-interleaving, large consecutive losses of data can be handled by low-power codes. This does not necessarily mean a reduction in required redundancy, but does give a vast reduction in the complexity of encoding and decoding. Block-interleaving will not help with large random errors, however, but can lead to greater correction capability than initially suggested by the power of the codes. A well-known cousin of the block-interleave is the **cross-interleave**, made famous by its use within the compact disc system. Cross-interleaving follows the same basic principles as the block-interleave, but it has a greater ability to spread errors and is also more complex to implement.

11

Multi-dimensional data structures

11.1 MULTI-DIMENSIONAL TIME DOMAIN CODING

In the examples so far the message block sizes have been determined by the size of the field used. If q is given by $q = 2^m$, where we are operating over GF(2^m), then each block has comprised $q - 1$, m bit symbols. There is, however, no reason why we can't use smaller or larger message block sizes within the same field if we so choose. In the single-error correcting example of Chapter 6, (6.1) and (6.2) were used to force two roots into the message vector. We can construct the message with only three data symbols if we want, as follows:

$$A + B + C + R + S = 0 \text{ and } A\alpha^1 + B\alpha^2 + C\alpha^3 + R\alpha^6 + S\alpha^7 = 0$$

R and S are calculated in exactly the same way as before. We have actually increased the message redundancy by doing this and so we can expect a payback in terms of error information. We can still only correct one symbol but we now have a higher probability of detecting (not correcting) more than one error. Previously, if only one of the syndromes was non-zero we could infer that more than one error had occurred. The probability of this detection is now greater since if k, the error index, points to 4 or 5, we again realize that more than a single error has occurred. In this case we won't attempt to correct the message. This is a shortened code.

A more interesting case is where we extend the block size beyond $q - 1$, since this actually yields an improvement in channel use (or an increase in the coding rate). Considering a single correcting code in the time domain, we can arrange the data in a two-dimensional manner, $q - 1$ symbols in each dimension. In the event of an error in this block we have three unknowns, the error pattern, the column index of the error and the row index. To solve these three unknowns we must add an extra (third) redundant symbol. This allows us to extend a message block over, say,

GF(2^3), to 49 symbols (46 data and three error check symbols) from seven. To do this, we set up the following conditions in the message, using the data $d_{i,j}$ ($1 \le i,j \le 7$).

$$\sum_{i=1,j=1}^{7} d_{i,j} = 0 \tag{11.1}$$

$$\sum_{i=1,j=1}^{7} d_{i,j} \alpha^i = 0 \tag{11.2}$$

$$\sum_{i=1,j=1}^{7} d_{i,j} \alpha^j = 0 \tag{11.3}$$

To satisfy these requirements three of the elements must be error check symbols, and their positions are not arbitrary. To locate and correct an error we must calculate three syndromes at the receiver, S_0 to S_2. In the event of a single error, say ξ, at position (i, j), S_0 will be equal to ξ, S_1 will be equal to $\xi \alpha^i$ and S_2 will be equal to $\xi \alpha^j$. To locate the error, we simply calculate $k_i = S_1/S_0$ and $k_j = S_2/S_0$. We have thus encoded 49 symbols (147) bits, using only GF(2^3). This idea can be extended to more dimensions over GF(2^3) by simply adding an extra symbol of redundancy per error per dimension.

Table 11.1 Comparison between 1D and 2D data structures

Comparison	Single error	Single error	Double error	Double error
	(a) GF(2^4)^2	(b) GF(2^7)	(c) GF(2^4)^2	(d) GF(2^7)
Message (bits)	900	889	900	889
Redundancy (bits)	12	14	24	28
Redundancy (%)	1.33	1.57	2.67	3.15
Correctable bits	4	7	8	14
Operations	705	381	1185	889
Op-bits	2820	2667	4740	6223
Op-elements	45 120	341 376	75 840	796 544

A more realistic example is to compare the field GF(2^4) when extended into two dimensions, with GF(2^7) in only one dimension since these represent similar message sizes. In Table 11.1, columns (a) and (c) comprise the two-dimensional case over GF(2^4), with a single and double correcting capability respectively, while columns (b) and (d) represent a one-dimensional example over GF(2^7), with a single and double symbol correction capability. The rows attempt to define some measures of comparison. The 'operations' row is an estimate of the number of operations required to calculate the syndromes for the message, while 'op-bits' scales this by the number of bits involved in each operation. The last row, 'op-elements' scales this again by the field size.

Whichever measure is most appropriate depends upon hardware implementation. For example, if a full-custom hard-wired solution was proposed then the bottom row would be a fair comparison. If, however, a microprocessor-type solution was to be used then a combination of operations and operation-bits would be more suitable. The four-bit solution over 16 elements blends easily with an eight-bit architecture which may also give it an advantage in some applications. Examination of the bottom row indicates that, for this measure, much less resource requirements are made by the smaller field. Even if we scale this row by correctable bits, we're still four to five times better off.

Because of the potential advantages that multi-dimensional correction strategies can offer in some implementations, we'll examine the solution to a single-error correcting code in GF(2^4) in three dimensions. The message size will be 13 500 bits (which may be unrealistically large), and the redundancy will be 0.12% with a four-bit (one-symbol) correction capability. We'll need one check symbol per dimension, plus one (for pattern), and in order to be able to satisfy the four requirements,

$$\sum_{i=1, j=1, k=1}^{15} d_{i,j,k} = 0 \ (S_0)$$

$$\sum_{i=1, j=1, k=1}^{15} d_{i,j,k}\alpha^i = 0 \ (S_1)$$

$$\sum_{i=1, j=1, k=1}^{15} d_{i,j,k}\alpha^j = 0 \ (S_2)$$

$$\sum_{i=1,\,j=1,\,k=1}^{15} d_{i,j,k}\alpha^k = 0 \ (S_3)$$

we must arrange for our check symbols to reside in different axes. The first check symbol we'll call p, and this is offset one position into the i axis, replacing data symbol $d_{14,15,15}$. The second check symbol, q, is offset into the j axis by one, replacing $d_{15,14,15}$, and the third symbol, r, is offset into the k axis by one, at position $d_{15,15,14}$. The last check symbol, s, will intersect all axes in the same place, replacing $d_{15,15,15}$. Over such a large field, generation of p, q, r and s based on an algebraic solution such as was used in Chapter 6 would be very difficult. Instead, we can use the erasure process discussed in Chapter 9. The four check symbols are initially set to zero and the syndromes are calculated as they would be in the receiver. This gives us four simultaneous equations:

$$S_0 = p + q + r + s$$
$$S_1 = p\alpha^{14} + q + r + s$$
$$S_2 = p + q\alpha^{14} + r + s$$
$$S_3 = p + q + r\alpha^{14} + s$$

Solving these gives

$$p = (S_0 + S_1)/(\alpha^{14} + 1)$$
$$= (S_0 + S_1)/(\alpha^3)$$
$$q = (S_0 + S_2)/(\alpha^3)$$
$$r = (S_0 + S_3)/(\alpha^3)$$
$$s = (S_0 + p + q + r)$$

In this solution the primitive polynomial $x^4 + x + 1 = 0$ was used to evaluate $(\alpha^{14} + 1)$ over $GF(2^4)$. At the receiver if any recalculated syndrome is non-zero, then the three-dimensional error position, i, j, k, is found from

$$i = S_1/S_0$$
$$j = S_2/S_0$$
$$k = S_3/S_0$$

and the error pattern, ξ, from

$$\xi = S_0$$

To correct the error we simply do the operation $d_{i,j,k} = d_{i,j,k} + S_0$. An example program performing this operation is included in Appendix D.

11.2 TWO-DIMENSIONAL FREQUENCY DOMAIN CODING

Following on from Section 11.1, the next question that immediately springs to mind is: can we perform a Fourier transform on a multi-dimensional data set? Double and triple correcting codes are still quite simple if we are prepared to use frequency domain processing. The answer to this question is yes, but only over two dimensions. The whole process is very similar to one-dimensional processing. The two-dimensional transforms can be effected by the following:

$$F_{i,j} = \sum_{x=0,y=0}^{q-2} d_{x,y}\alpha^{(i+y)(j+x)}$$

$$d_{i,j} = \sum_{x=0,y=0}^{q-2} F_{x,y}\alpha^{-(i+y)(j+x)}$$

where $q = 2^m$, over the field GF(2^m). Notice that if j and y are set to zero we end up with the one-dimensional transform. To illustrate, we'll use a small example over the manageable GF(2^3). Table 11.2 shows a 49-symbol message arranged into a two-dimensional matrix, with three zeros (bold) inserted into positions (0, 0), (0, 1) and (1, 0). These are sufficient to give a single-symbol error correction capability to the message.

Table 11.2 Two-dimensional message over GF(2^3)

0	**0**	0	6	6	7	4
0	2	2	5	2	3	0
6	3	7	2	3	7	4
1	0	2	1	1	6	0
2	3	6	2	6	0	6
5	0	2	3	2	1	5
2	6	5	1	6	1	6

Table 11.3 shows the message in the time domain after inverse Fourier transforming. The boxed element will be corrupted upon arrival at the receiver.

Table 11.3 Time domain form of the message

0	6	7	3	5	3	6
4	1	5	7	0	7	5
3	5	2	2	1	5	7
0	2	6	7	0	5	2
7	2	4	4	6	1	2
1	1	1	0	3	7	3
2	2	1	0	4	2	3

If we now add the error 6 to the message (Table 11.3), in location (4, 5) and Fourier transform, we get the following in Table 11.4. The Fourier transform of the error itself is given in Table 11.5.

Table 11.4 Corrupted frequency domain message

3	1	6	4	1	3	1
2	7	4	1	3	4	3
3	4	1	1	7	5	5
7	6	4	7	7	0	6
6	0	0	5	3	1	4
4	4	4	6	0	2	2
5	4	3	0	5	4	2

Table 11.5 Frequency domain form of the error

3	1	6	2	7	4	5
2	5	6	4	1	7	3
5	7	6	3	4	2	1
6	6	6	6	6	6	6
4	3	6	7	5	1	2
1	4	6	5	2	3	7
7	2	6	1	3	5	4

From Tables 11.4 and 11.5 you will notice that wherever there is a zero in the original message, the corrupted message (Table 11.4) now shares a common value with the error spectrum. Because three zeros were deliberately placed in the top left corner of the message, we are guaranteed of knowing at least three error spectra from the received and corrupted message. Looking at the error (Table 11.5) you'll notice a complete row and column of 6s and it's tempting to think that their intersection point is the error location. We know, however, that the error was placed at (4, 5) and not (2, 3).

Using an REC for 1 error we can reconstruct the top row of the error spectrum based upon the two known error spectra, 3 and 1. For a single error, $L = E_1/E_0$. This gives us $Lx_0 = \alpha^7/\alpha^3$ or α^4. Using this feedback multiplier and preloading the REC with 3, the sequence α^{3+4x} is generated, where $(0 \leq x \leq 6)$ which is

$$3, 1, 6, 2, 7, 4, 5$$

We also know two error spectra in the first column, 3 and 2. Using these in exactly the same way as above, $Ly_0 = \alpha^1/\alpha^3$ or α^5. This gives the sequence α^{3+5y} where $(0 \leq y \leq 6)$ or

$$3, 2, 5, 6, 4, 1, 7$$

which is the first column of the error spectrum. At this point you may have realized that we don't, in this simple case, need to complete the error spectrum in order to identify the error. For the case of a single error the multiplier L points to its location which from (Lx, Ly) is $(4, 5)$. From the Fourier transform we also know that message element $(0, 0)$ is given by $\xi \alpha^{ij}$ where ξ is the error pattern, and i and j its x–y position. From this

$$\alpha^3 = \xi \alpha^{4.5}$$
$$\xi = \alpha^3/\alpha^6 = \alpha^4 \ (6)$$

For a bit of practice we'll complete a little more of the error spectrum anyway since this demonstrates an operation which can be extended to many errors. Having completed the first row and column of the error spectrum using the three known values in the top left corner of the message, we now know at least one spectral component per line and column. If we already know the value of L then this is sufficient to set up the REC for each row or column. We can choose at this point whether to reconstruct the spectrum by rows or columns. Choosing rows for this example we already know that $Lx_0 = \alpha^4$.

By moving one row down the matrix we have modified our calculation point with respect to the error position by one. This is reflected in L as an increase of α, so $Lx_1 = \alpha Lx_0 = \alpha^5$. Using the known spectral value of the second row, 2 (α^1), we create the sequence α^{1+5x} which is

$$2, 5, 6, 4, 1, 7, 3$$

and so on. We could complete the spectrum by columns in exactly the same way, incrementing Ly for each step to the right. Taking the second column as an example $Ly_1 = \alpha^6$ and using the known spectral component 1 (α^7), we have the sequence α^{7+6y} which is

$$1, 5, 7, 6, 3, 4, 2$$

Extending the two-dimensional transform to more errors requires a little consideration of the problem. The REC for two errors has two multipliers, L_1 and L_2. To find these in either direction (x or y), we need to know four spectral components in a row. Once we know L_1 and L_2, we need a further two components to reconstruct any other row. To this end the double-correcting two-dimensional message is set up as shown in Table 11.6.

Table 11.6 Creating a double-correcting message

0	0	0	0	6	7	4
0	0	2	5	2	3	0
0	3	7	2	3	7	4
0	0	2	1	1	6	0
2	3	6	2	6	0	6
5	0	2	3	2	1	5
2	6	5	1	6	1	6

Table 11.7 Time domain (encoded) message before corruption

0	6	0	6	4	3	2
5	7	7	4	5	0	1
1	6	0	4	3	3	6
0	4	0	3	6	6	3
5	1	5	7	5	7	2
4	0	7	3	2	1	5
0	3	7	4	7	6	7

In Table 11.6 the four added zeros (bold) on the first row and column allow initial calculation of the two multipliers in either direction since in the event of errors we will know four adjacent spectral components in both horizontal and vertical directions. At position (1, 1) another zero means that we know two spectral components in both the second row and column. We can therefore complete both a second row and column using multipliers calculated from the first row and column. Now we will have the multipliers in both directions and at least two known error spectra for every row (or column depending on which we choose to solve) which means that we can complete the error spectrum.

Table 11.7 shows the time domain message after inverse Fourier transforming, and Table 11.8 shows the decoded (frequency domain) message after the addition of two errors, 6 at (4, 5) and 5 at (2, 2). The error spectrum is also shown in Table 11.9 for comparison.

Table 11.8 Reconstructed message with errors

0	6	7	6	7	6	3	
5	7	2	0	0	1	7	
4	2	3	6	3	0	6	
2	7	3	4	5	5	0	
0	3	4	7	2	4	3	
1	1	1	3	5	7	7	
7	0	0	6	2	1	3	

Table 11.9 Error spectrum

0	6	7	6	1	1	7	
5	7	0	5	2	2	7	
4	1	4	4	0	7	2	
2	7	1	5	4	3	0	
2	0	2	5	4	4	5	
4	1	3	0	7	6	2	
5	6	5	7	4	0	5	

From earlier examples we know that

$$L_1 = \left(E_0 E_3 + E_1 E_2\right)\Big/\left(E_1^2 + E_0 E_2\right)$$
$$L_2 = \left(E_1 E_3 + E_2^2\right)\Big/\left(E_1^2 + E_0 E_2\right)$$

Working across the top row of the received message,

$$L_1 x_0 = \left(0 + \alpha^4 \alpha^5\right)\Big/\left(\alpha^4 \alpha^4 + 0\right) = \alpha$$

$$L_2 x_0 = \left(\alpha^4 \alpha^4 + \alpha^5 \alpha^5\right)\Big/\left(\alpha^4 \alpha^4 + 0\right) = \alpha^6$$

Preloading the REC with 0 and 6 generates the sequence

$$0, 6, 7, 6, 1, 1, 7$$

the top row of the error spectrum. Doing the same but vertically, $L_1 y_0 = \alpha^3$ and $L_2 y_0 = \alpha^7$. This time we generate the first column after preloading with 0 and 5:

$$0, 5, 4, 2, 2, 4, 5$$

In order to complete any more rows or columns we have to understand how to modify L_1 and L_2 as we progress across or down the message. For each move down by one row, $L_1 x$ is multiplied by α, just as in the previous example. $L_2 x$, on the other hand, increases by α^2. Recalling the one-dimensional situation where two errors are located at positions i and j, it was demonstrated that $L_1 = \alpha^i + \alpha^j$ while $L_2 = \alpha^{i+j}$. If we move one then relative to our last calculation, i and j increase by one so now

$$L_1 = \alpha^{i+1} + \alpha^{j+1} = \alpha(\alpha^i + \alpha^j)$$

and

$$L_2 = \alpha^{i+1+j+1} = \alpha^2 \alpha^{i+j}$$

or α and α^2 times their previous values. $L_1 x_1 = \alpha^2$ and $L_2 x_1 = \alpha$ in this example and preloading with 5 and 7 gives the sequence

$$5, 7, 0, 5, 2, 2, 7$$

or the second row of the error spectrum.

Table 11.10 Time domain error pattern

0	0	0	0	0	0	0
0	0	0	0	0	0	0
0	0	5	0	0	0	0
0	0	0	0	0	0	0
0	0	0	0	0	0	0
0	0	0	0	6	0	0
0	0	0	0	0	0	0

Table 11.10 shows the time domain error pattern after inverse Fourier transforming the completed spectrum. Table 11.11 shows one possible ordering of the completion of the error spectrum. Also shown in the table are the multiplier values for each row and column.

Table 11.11 Reconstructing the error spectrum

		x	0	1	2	3	4	5	6
		L_1	α^3	α^4	α^5	α^6	α^7	α^1	α^2
y	L_1	L_2	α^7	α^2	α^4	α^6	α^1	α^3	α^5
0	α^1	α^6	1	$\rightarrow$					
1	α^2	α^1	2	3	$\rightarrow$				
2	α^3	α^3	$\downarrow$	4	5	6	7	8	9
3	α^4	α^5	$\downarrow$	$\downarrow$	$\downarrow$	$\downarrow$	$\downarrow$	$\downarrow$	
4	α^5	α^7							
5	α^6	α^2							
6	α^7	α^4							

In order to understand how to modify the multipliers for greater correcting capacity, consider the origin of L_{1-3} for a three-symbol correcting code over a one-dimensional message. If the three error positions are at i, j and k then from

$$\left(\alpha^x + \alpha^i\right)\left(\alpha^x + \alpha^j\right)\left(\alpha^x + \alpha^k\right) = 0$$

we derive the expression

$$\alpha^{3x} + L_1\alpha^{2x} + L_2\alpha^x + L_3 = 0$$

so

$$L_1 = \alpha^i + \alpha^j + \alpha^k$$

and increases by α per row or column change

$$L_2 = \alpha^{i+j} + \alpha^{i+k} + \alpha^{j+k}$$

so increases by α^2 per row or column change

$$L_3 = \alpha^{i+j+k}$$

so increases by α^3 per row or column change.

It should be fairly easy to see the pattern that is emerging now. Appendix E demonstrates the previous example where two errors are added to a 49-symbol message.

11.3 PUNCTURED CODES

It was demonstrated in Section 7.1 that by using a process called erasure, a single-symbol correcting code can be used to correct two errors if their position within the message is known. This concept was used again in Section 9.4 in order to complete the Fourier transform of an error spectrum using a knowledge of where the errors were, and only half the normally required known error spectra. If we denote symbol errors as s and errors with known position, corrected by erasure, e, then for a t symbol correcting code

$$2t \geq 2s + e$$

In systems where channel noise is subject to large variations (typical in mobile communications systems), erasure decoding can be used to good effect in a number of ways. First, the receiver can monitor the received signal level or power and, in the event of fades, declare a symbol as erased. The symbol may have been received correctly and it may not. If the fading threshold is set appropriately then there is a good chance that the symbol would have been in error. By declaring the symbol erased, the receiver tells the decoder where the error is. The decoder does not have to locate the error, simply setting the symbol to zero. In this way potentially up to twice the number of errors can be corrected.

Punctured codes exploit this property to improve the throughput on such variable quality channels. If a code has to be arranged so as to cope with the worst possible channel conditions, then for much of the time far more bandwidth is being used for error control than is necessary. This scenario is partly managed by an automatic repeat request protocol where the receiver requests retransmission of a packet or message in the event of too many errors. The redundancy can now be set for more typical channel conditions making better use of the available bandwidth for data. In the event of a retransmit request, to simply retransmit the same message again, however, is likely to achieve a similar failed result, especially if the channel fade persists.

If a message is constructed such that over half of it is redundancy and it is then divided into two equal parts, it is possible to reconstruct the original message from either of the halves. The missing half of the message forms known erasures of which we can correct twice what would normally be expected. If only one half of the message is transmitted, the majority of its capacity is used to replace the missing half by erasure decoding. Any remaining capacity can be used to correct real channel errors. In the event of transmission failure and a retransmit request, the other half of the message is transmitted. At this point, the second half of the message will be decoded independent of the first, again having minimal capacity to correct real errors. So far we are little better off than if the original message had been retransmitted and, if the channel fade persists, decoding failure results again. If the second half does not decode then the two halves of the message are brought together to create a complete message with massive correcting capability, hopefully sufficient to overcome the fade. If failure results again, the first half of the message will be retransmitted and so forth.

11.4 DISCUSSION

In this section we have examined block-interleaving and mechanisms for extending the message sizes of small fields. Interleaving has the effect of decorrelating large errors, converting them into lots of small but manageable errors. Block-interleaving is typically augmented by the addition of inner codes which while easily overpowered by large errors, weed out small random errors which might otherwise compromise the operation of the outer codes over which the large errors are distributed. The advantage of block-interleaving is a very powerful correctional capability with the use of only relatively humble correcting codes. In terms of redundancy we may well be no better off, but the computation can be orders of magnitude simpler. The cost of block-interleaving is a small increase in decoding time. Because Reed–Solomon coding is based around blocks little extra hardware is necessary to support the interleave.

A surprise additional bonus of interleaving is the interaction of the inner and outer codes. Where the capacities of both inner and outer codes have been exceeded, the knowledge accrued from the intersection of the inner and outer syndromes allows us to mount a feasibly sized trial-and-error search for the errors, in many cases leading to complete correction. Time domain message expansion into multi-dimensional messages, and frequency domain expansion into two-dimensional structures allows us to create realistically sized messages using only small fields. Where hardware processing is to be used (as opposed to a software-based solution) this can provide around a four to five times reduction in the hardware complexity. Where only modest error correction is required, we also see a reduction in the redundancy and so an improvement in channel use.

Last, punctured codes were considered as a means of making good use of bandwidth in situations where very variable signal quality is expected. These operate on the principle that we can correct twice as many erasure errors as unknown errors.

12

Convolutional coding

There are two main competing error control strategies and so far we have only looked at block codes. As their name suggests, with block codes the data are compiled into blocks prior to encoding and transmission and must be subsequently recompiled into blocks for decoding and error correction. The process is fairly mechanistic and the correction capacity very predictable. The second main strategy is called convolutional or **trellis** coding. Here data may be error encoded on a continuous basis without the need for compilation into blocks. This may seem a curious place in which to introduce convolutional coding and the justification for this is that its most major impact is upon packaging. The majority of the book has been about block coding of one sort or another and it would be a crime to give no mention of the alternative.

Convolutional encoding circuitry is very trivial and as a result this technique may find its way into applications where power consumption is a very important issue. Unfortunately, decoding and error correction is not trivial, neither is it very mechanistic. There are several strategies which may be employed to perform decoding and each one is a balance of compromises such as complexity and decoding latency. Some of the key features of this kind of error coding are its easy compatibility with soft decision decoding and an ability to modulate raw data directly into a usable channel code (suitable for launching onto a transmission medium).

Figure 12.1 represents a typical convolutional encoder. The **coding rate**, or ratio of input bits to output bits, is ½ for this example since for every bit input, two bits will be output. The circuit also has a **constraint length** of three. The constraint length is a measure of how many bits effect the current output, or for how many clocks a given bit will persist in the output. By inputting more than a single bit at a time, coding rates of greater than ½ are possible, but first we'll examine the operation of this circuit.

This kind of coding is best visualized by generating a trellis diagram which reflects its operation. For the above circuit the trellis appears as shown in Fig. 12.2.

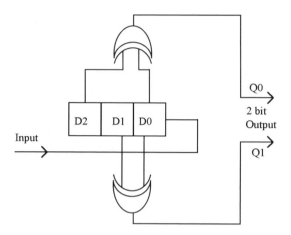

Fig. 12.1 A convolutional encoder.

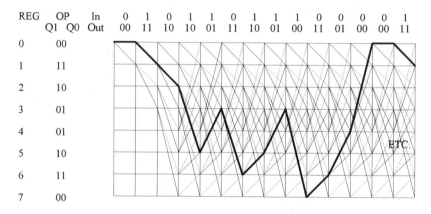

Fig. 12.2 Trellis diagram for a convolutional encoder.

The encoder has three registers and therefore can reflect eight states. These are labelled downwards in the trellis from 0 to 7. The number of register bits determines the **depth** of the trellis. On the horizontal axis are input/output bits or time. By means of the XOR gates the three register bits are converted into two output bits Q_0 and Q_1, sometimes defined in the polynomial form $Q_0 = 1 + x^2$ and $Q_1 = 1 + x$. These are shown in the second column and in a system would be multiplexed together to form a single bit-stream at twice the input bit rate. Inevitably the output bit patterns are repeated down the column since in this case two bits are being used to convey eight states.

Each time a new bit is input to the three-bit register the oldest bit falls out the end and the state of the registers potentially changes. This is plotted on the trellis with narrow solid lines representing an input 1 and dotted lines representing an input 0. From any one of the eight states we can move through the trellis to two new valid states depending on the input. Starting from 0, the bold line represents the path carved through the trellis by the input data given on the top line. Below the input data is the output sequence with two bits for every one input, and this is what is transmitted.

The circuit has imposed a constraint on the way in which the transmitted sequence pairs can be generated and changed. At the receiver the decoder plots the progress of the data through the trellis, comparing each bit pair to what is expected based on the current position in the trellis. Where errors are found the decoder attempts a kind of 'best fit', matching the received bit-stream to the closest valid path through the trellis. This closest path is deemed to be the corrected data.

This is where life gets a little tricky and there exist several strategies for performing this 'best fit'. Clearly one could compare the received sequence to every possible path through the trellis and evaluate a cost for each path, finally choosing the least expensive path. This rather defeats the object of convolutional coders, however, since we must wait until the entire message is received before any result can be output. Not only this but the number of possible paths would quickly become unmanageable. The **Viterbi** algorithm, named after its inventor, makes a simplification to the decoding process using the following idea.

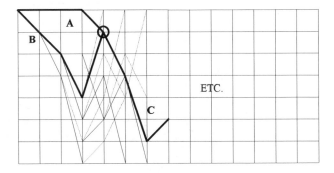

Fig. 12.3 Optimizing using the Viterbi algorithm.

Considering the trellis in Fig. 12.3, there are two routes (A and B) to the node that is circled, and these are shown in bold. Routes A and B will have a cost which is based upon the difference between them and the received sequence from the transmitter. Two routes thus meet with two associated costs. All routes emanating from the circled node will inherit the costs of A and B and will incur the same new costs from this point on. C is one such route and if the combined cost of route AC is smaller than the combined cost of route BC, then there is no point in continuing with B since it can never catch up with A. What this means is that we only need to keep a running track of 2^n routes where there are n registers.

12.1 FINDING THE CORRECT PATH

If we corrupt the data stream given previously from

$$00, 11, 10, 10, 01, 11, 10, 01, 00, 11, 01, 00, 00, 11$$

to

$$00, 11, 10, \mathbf{11}, 01, 11, \mathbf{00}, 01, 00, 11, 01, 00, 00, 11$$

we can attempt correction by constructing a cost table in the following way. We have enough vertical positions to accommodate the eight paths that we will allow, plus one for the incoming data stream at the top. Table 12.1, shown below, is constructed on the basis of the encoder of Fig. 12.1.

Table 12.1 Finding the correct path through the trellis

	00	11	10	11	01	11	00	01	00	11	01	00	00
0/00	0	2	3	5/7	6/2	4/4	4/4	5/3	3/3	5/7	6/6	6/2	2/6
1/11	2	0	3	3/5	6/2	2/2	6/6	5/3	5/5	3/5	6/6	8/4	4/8
2/10		3	0	4/4	5/3	3/5	3/5	8/4	4/6	6/6	5/7	7/5	5/7
3/01		3	2	4/4	3/1	3/5	3/5	6/2	4/6	6/6	3/5	7/5	5/7
4/01			5	1/5	4/2	4/6	4/2	3/5	5/5	5/5	6/2	6/6	6/6
5/10			3	1/5	6/4	4/6	4/2	5/7	5/5	5/5	8/4	6/6	6/6
6/11			4	2/4	5/5	1/5	5/5	4/4	4/6	4/2	7/5	5/7	7/5
7/00			4	4/6	5/5	3/7	3/3	4/4	2/4	6/4	7/5	3/5	5/3

Starting from 0, the incoming bit pair 00 is compared to what we could expect at this point. From the trellis we would expect 00 or 11. If the pair

where we can

go from the 0 state

we can go to 0 state   or   1 state => 00   or   11

should have been 00 then the cost is nothing since it was 00, and we can place 0 in the corresponding position (light shading) in the table. For the received sequence to have been 11, two bits would have had to be in error so the cost of this selection is 2, which is placed in the table appropriately (darker shading). We now have four paths to consider, emanating from the two previous possibles, and must count the costs of each, filling in the table as we go. In the next column there are eight paths while in the next column, paths meet back together. This is reflected in that this and subsequent columns contain two values. Also in this column the first bit error is evident in that there is no remaining path with a cost of 0, and two with a cost of 1. At this point the Viterbi algorithm kicks in and we have to decide which paths to discard, keeping those with the lowest cost. Where the two paths have a similar cost we can never know which is the correct one. The selected paths are shown bold in each column.

It is tempting to look at the table and pick out the corrected path as we go along, but this is not possible for the following reasons. To start with, where the first error occurs two paths have the same cost, and second, if both bits of a pair were inverted then the error would result in a cost of 0 in an incorrect path and a cost of 2 in the correct one, not showing up until later. In other words, we can't be sure which is the correct track until some time after an error(s) has occurred.

When a route displays a constant cost over a number of bits it is possible to start working backwards to find out from where it came. In this example one route exhibits a cost of two bit errors for the last seven columns and, in the absence of more errors, this would have continued. Picking up this route and working backwards allows us to select the least-cost path even though some columns share the same minimum cost between two routes. This is the case in both columns where the bit errors occur, but it does not stop us finding the correct path.

The bit costs in the table above have been calculated in whole numbers of bits but it is a simple extension of this to include the probabilities generated by soft decision decoders. Almost certainly this will eliminate the chance of paths sharing the same costs, and in fact does enhance the correction capability of the code. Theoretically, soft decision decoding can provide an improvement of some 3 dB in a communications system, but in practice the result is more like 2 dB. This discrepency may stem in part from the fact that a linear quantization is usually used to generate the bit probabilities. The ideal quantizations should, in fact, be slightly non-linear (0, 1, 2, 3, 4, 5, 6 and 8.67).

12.2 OTHER CONVOLUTIONAL CODERS

At first glance, this kind of encoding might appear to limit us to coding rates if $1/n$ where n is an integer. By providing more inputs, coding rates closer to one can be realized. Such an encoder could appear as follows in Fig. 12.4, having a coding rate of $2/3$.

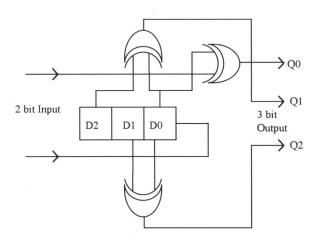

Fig. 12.4 A convolutional encoder with a coding rate $2/3$.

As with the previous encoder this one has what in signal processing terms would be called a finite impulse response. This simply means that the effect of any input to the circuit will only persist in the output for a finite duration or number of clocks, after which it drops out of the end of the shift register. Such encoders can be defined by their impulse response which for Fig. 12.1 is (Q_1/Q_0) 111001, for a single 1 input. There is no reason why these encoders can't include feedback, much as the cyclic redundancy checker has. The signal processing equivalent would have an infinite impulse response which means that once input to the circuit, any bit will potentially affect the output for the remainder of the message. Circuits of this kind are more generally known as scramblers, and are used to shape the AC and DC characteristics of raw binary data before launching it onto a communications medium. If we can throw in an error correction capability as well, then convolutional coding presents a very economic solution to some applications. This kind of modulation is known as trellis-coded modulation (TCM).

12.3 DISCUSSION

Convolutional coders are clearly attractive as far as hardware implementation of the encoder goes but the complexities of decoding may offset this in certain cases. In practical systems a trellis depth of about 64 probably represents the maximum that is manageable since this implies simultaneous evaluation of 64 cost paths.

This kind of coding is ideally suited to smoothly distributed random errors typical on communications channels. What convolutional coding will not do well is cope with large losses of consecutive data such as we might encounter in a digital storage system. Of course, block-interleaving could be added to decorrelate these errors but this would entail compiling data into blocks for which block codes are much better suited. The simplicity of the encoding would, therefore, be compromised and an undesirable latency added into the channel. Even so, there are so called 'helical' interleavers which may be used with convolutional coders to good effect.

There is no reason why convolutional codes cannot form the inner code of a cross-interleaved message. It has been noted that convolutional inner codes and Reed–Solomon outer codes in an interleaved message form an ideal partnership. Convolutional codes deal well with random errors, but where their capacity is exceeded the output of the decoder represents a burst error, handled admirably by the outer block codes. Also, no convenient way has yet been found of incorporating soft decision decoding directly into block codes. Where the extra 2 dB of coding gain is essential there is, therefore, no alternative but to use a convolutional inner coder.

This combination of inner convolutional codes and outer Reed–Solomon codes formed the basis of the communications link used by the Voyager expeditions to Uranus and Neptune. Deep-space communication suffers mainly from random errors for which the convolutional code will recover well, but by augmenting the codes with an outer Reed–Solomon code, huge coding gains are achieved. The Galileo mission to Jupiter almost failed due to the refusal of a high gain antenna to open. This meant all communication had to take place using a low gain antenna resulting in a drastic reduction in the possible data rate. This event prompted a large effort to increase the coding gains of the codes used. The result is a decoding system that works beyond what was previously thought possible, demonstrating the synergy created by interleaving. The decoder uses both soft decisions and erasure to achieve this enhancement and in some cases, the natural redundancy of the space images themselves.

12.4 SYMBOL-BASED CONVOLUTIONAL CODING

Where we can afford a coding rate of ½, it is possible to conceive of coding schemes based around finite fields which allow encoding and decoding without the need for compilation into large blocks. If we take a sequence of data, say, d, we can generate a second redundant sequence, r, as follows:

$$r_i = \sum_{j=-k}^{k} d_{i+j}$$

where d_{i+j} is deemed to be 0 outside the range $(0 \le i + j < n)$ for an n symbol message; k sets the impulse width of the coder from $I_w = 2k + 1$. Effectively we are convolving the data with a square function. Clearly the encoding latency will be k symbols since we can't generate r_i until d_{i+k} has arrived. Consider the data $d_{0-10} = 0, 2, 7, 3, 6, 1, 7, 0, 0, 1, 2$ and $k = 1$. The redundant message r is calculated in Table 12.2.

Table 12.2 Calculating *r* from *d*

i	0	1	2	3	4	5	6	7	8	9	10
d_i	0	2	7	3	6	1	7	0	0	1	2
r_i	2	5	6	2	4	0	6	7	1	3	3

If we interleave d and r then the transmitted message is

$$0, 2, 2, 5, 7, 6, 3, 2, 6, 4, 1, 0, 7, 6, 0, 7, 0, 1, 1, 3, 2, 3$$

I have dubbed this *interleaved data and redundant convolutional coding* (IDARCC), for obvious reasons. This method of coding has the advantages of time domain data, largely visible even in the presence of uncorrectable errors, and the advantage (in some applications) of small latency, having no need to construct large blocks. For the purpose of decoding these messages and performing error correction it's easiest to start by considering the effects that errors have on them. We'll begin by examining the case where a single symbol is received in error.

Two sorts of single-symbol error may occur and their effect upon the message is profoundly different allowing us to differentiate between them. First, an error may affect a data symbol, and second, it may corrupt one of the redundant symbols. To check for errors at the receiver, two syndromes are calculated. The message is de-interleaved to give two symbol streams d' and r'. d' is re-encoded as at the transmitter and the resulting coded stream is added to r' to create syndrome stream S_0. r' is decoded to regenerate d and this symbol stream is added to d' to create syndrome stream S_1. This is explained in Fig. 12.5.

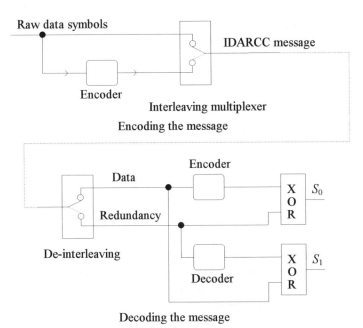

Fig. 12.5 **Generating syndromes S_0 and S_1.**

Decoding of r is only possible if the first k symbols of any message are known. This can be arranged by setting them to zero at the encoder, reducing the coding rate very slightly from ½. Figure 12.6 illustrates the coding of r_0, and from this it is clear how d_1 is calculated from r_0. Once d_1 is known, d_2 can be found as r_1 arrives and so on.

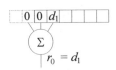

Fig. 12.6 **Calculating d from r.**

As the message arrives, the syndrome streams S_0 and S_1 are generated, and these will be zero in the absence of errors. If an error occurs which affects only a data symbol then S_0 will exhibit $2k + 1$ burst of non-zero values equal to and centred on the error. S_1, on the other hand, will experience only a single non-zero syndrome equal to and at the error. An error in a redundant symbol, however, will cause only a single non-zero symbol in S_0 equal to and at the error and a repetitive double symbol pattern in S_1 which starts k symbols after the error, and repeats every I_w symbols thereafter. It is easy to demonstrate these two scenarios by corrupting the message generated previously to

$$0, 2, 2, 5, \underline{\mathbf{6}}, 6, 3, 2, 6, \underline{\mathbf{0}}, 1, 0, 7, 6, 0, 7, 0, 1, 1, 3, 2, 3$$

The syndrome sequences are

$$S_0 = 0, 1, 1, 1, 4, 0, 0, 0, 0, 0, 0$$
$$S_1 = 0, 0, 1, 0, 0, 4, 4, 0, 4, 4, 0$$

Since $k = 1$ in this case, the impulse width of the system is three symbols $(2k + 1)$, demonstrated by the first error (in d'). The second error was in r', giving the cyclic double symbol burst in S_1. Like encoding, the decoding process operates over a window of I_w symbols. Correction of single symbol errors (separated by at least I_w error-free symbols) is trivial, but as with traditional convolutional coding, errors can't be corrected until sometime after they have occurred. At the very least, evaluation of S will require a latency of k symbols. Where the isolated single non-zero syndrome is discovered in S_0, it belongs to r_4 and in this case can be corrected by adding its current value to the non-zero syndrome. If only a single symbol of d' has been corrupted, as in the first error above, then the central non-zero symbol of the S_0 burst equals the error pattern. d_2 is therefore $6 \oplus 1 = 7$.

Where isolated errors occur, decoding latency remains at k data symbols since we can correct them as soon as the syndromes are available. Where errors become more complicated, such as in the case of a burst error, the syndromes additively combine if overlap occurs. Consider the following syndrome generated from a message with $k = 2$.

$$S_0 = 0, 4, 4, 7, 7, 7, 3, 3, 0, 0, 0, 0, 0, 0, 0, 0$$
$$S_1 = 0, 0, 0, 4, 0, 3, 0, 0, 0, 0, 0, 0, 0, 0, 0, 0$$

Since there is no repetitive syndrome in S_1, both errors must have occurred in d'. S_1 tells us exactly what the errors are, but the impulses in S_0 overlap. Where the five 4s and five 3s overlap, we get the combined symbol which is $100 \oplus 011 = 111$ or 7. The same thing happens when errors occur in r'. The following syndromes are for $k = 2$, and two adjacent corrupted r's.

$$S_0 = 0, 0, 0, 4, 3, 0, 0, 0, 0, 0, 0, 0, 0, 0, 0, 0$$
$$S_1 = 0, 0, 0, 0, 0, 4, 7, 3, 0, 0, 4, 7, 3, 0, 0, 4$$

The errors were 4 and 3, so we would expect a cyclic 44 pair and a cyclic 33 pair. Again, they overlap so we see the combined syndrome 7. As long as we know at least one non-overlapped symbol, say at the start of a sequence of non-zero syndromes, then we can calculate the rest. If the second 3 in S_1 had been obscured by further errors in r, we could have found it from $4 \oplus 7 = 3$. The cyclic nature of the S_1 syndrome is the key to the success of this very simple code. Where gross errors occur, the combination of the syndromes can make identification of errors very difficult. However, because errors in r ripple on indefinitely through S_1 (at least until we correct them) after the error has occurred, we can examine these syndromes long after the burst error, when we are sure that no corruption of S_1 is taking place due to errors in d'. We can find a suitable error-free stretch of code by monitoring S_0.

To test out the theory, the following syndromes were generated by the burst error:

$$\xi = 60F2001CF13CF709CE8AFCBEFB$$

starting with 6 in d'_7 plus two random errors. The symbol size is four bits and $k = 7$. The burst error is 26 symbols long, representing a possible loss of up to 104 successive bits.

S_0 = 699874BB5FCA83DC71280FF3B4F00000044444444448444440000000000000
S_1 = 00000006F01F3F0EA362BE74625B0022CDDBE74665B0022CDD7274625B0

If we take a sample of I_w symbols S_1, from a region where S_0 suggests that there are no further errors, then we can attempt to decipher it. Since isolated random errors are easily dealt with, we don't require a completely error-free window as in this case. The simple nature of the syndromes towards the end of the list allows us to make a simple adjustment to S_1. First, we'll take a sample of I_w symbols from S_1, starting at 0022:

0022CDDBE746**6**5B

Examination of the impulse in S_0 indicates first that the error 4 has occurred in d', coinciding with the bold, underlined symbol in S_1 above. This must be changed to $6 \oplus 4 = 2$ giving

0022CDDBE746**2**5B

The isolated 8 in S_0 indicates a single error in r', but this will not become manifest in S_1 for a further seven symbols (and every I_w symbols after), well outside our selected window. Because we are only tackling S_1 we are only dealing with errors in r' at the moment. Table 12.3 lists the decoding process.

Table 12.3 Recovering errors in r' from S_1

i	28	29	30	31	32	33	34	35	36	37	38	39	40	41	42
S_{1i}	0	0	2	2	C	D	D	B	E	7	4	6	2	5	B
Er_{i-1}	0	0	0	2	0	C	1	C	7	9	E	A	C	E	B
Er_i	0	0	2	0	C	1	C	7	9	E	A	C	E	B	0

The bottom row of Table 12.3 represents the XOR of the middle two rows, while the lower of the middle two rows represents the bottom row shifted right by one symbol. Careful examination of the bottom row reveals it to be the burst error in r', or every alternate symbol in the burst error. We now have to calculate the index of the errors. First, we must subtract k. The syndrome in S_1 will always be displaced by at least k symbols to the right of the error position, but it may also be displaced by a further integer multiple of I_w symbols after that because the syndrome repeats. Depending on the complexity of the error, this may be obvious as in this example.

Considering the error 2, at $i = 30$, our two possibilities for its location in r' are r_{23}, or r_8, i.e. $30 - k$ or $30 - k - I_w$. If the errors started at r_{23}, with a burst of 12, we would still be seeing evidence of them in S_0. In this case, the burst must start at r_8. XORing r' with E_r, calculated above, having aligned them appropriately, we can now recalculate the two syndromes, giving

S_0 = 699874BB7F0B4442DDC30FF3B4F00000044444444484444000000000000000
S_1 = 00000006F01F3F0C8FBF00000000000000000004000000000CC00000000000

This time we see no evidence of any cyclic errors in S_1. We need to beware only to correct parts of the message at a time as we go along; this is not a block code, but must be corrected on a continuous basis. The CC towards the end of S_1 is a cyclic error, we just don't know it yet, so we might not attempt correction here until more of the message has arrived although in this case S_0 gives us a strong clue as to the nature of the error. The remainder of S_1, however, represents the error in d'. You can check this by referring back to ξ. To correct this portion of the message, d' is simply XORed with or added to S_1.

This code is powerful enough to correct $4k - 2$ successive message symbol errors, provided a 'reasonably' error-free window of I_w symbols follows the burst. 'Reasonable', will clearly be the subject of some debate. If soft decisions are available, or if the message can provide some context, then $4k$ successive message symbol errors are correctable, although there will be several possible solutions. The Pascal code for this example is listed on the web, and since only a summation of symbols is required, it does not require the support of other libraries.

Decoding latency will be typically k data symbols for no errors or random errors. Where large bursts occur, we will be forced to wait for a 'quiet' period in order to effect correction. The decoding latency is thus variable and will be shaped by k which in its turn determines the scope of the burst errors that will be correctable. We can provide a trade-off between latency and burst error capability. Encoding is as trivial as normal convolutional coding, and decoding is done algebraically unlike convolutional coding which is a kind of best-fit search. In my opinion this kind of mixed coding provides a viable alternative to convolutional coding in situations such as digital phones where latency often precludes block coding. Coding rates are similar to convolutional coders although the latency will probably be somewhat greater. Decoding will be simpler than traditional convolutional codes, however, and burst error capability is programmable.

Figure 12.7 shows how an encoder can be realized using less hardware than might initially be expected. Normally a parallel summation of n values requires an n input parallel adder (for parallel operation). However, because our input is sequential, we need only add the first entering symbol, subtract the last (add one) exiting symbol (which is also adding!) and accumulate the sum. In this way, regardless of k, we need only three inputs to the summer. Because the symbol size is arbitrary, we have a range of possible solutions for a given error capability. Suppose we want to protect against burst errors of up to 50 bits. With symbol bit widths of m, then $m(4k - 2) \geq 50$. Table 12.4 lists some possible solutions.

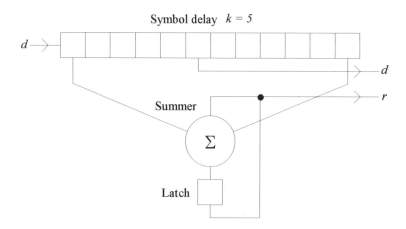

Fig. 12.7 Redundant symbol encoder using feedback.

Table 12.4 Solutions for m and k for 50-bit burst error capability

m	k	Burst (bits)
1	13	50
2	7	52
3	5	54
4	4	56
5	3	50
6	3	60
7	3	70
8	3	80
9	2	54
10	2	60

Clearly some combinations such as $m = 1$ and $m = 5$ give exact and, therefore, most economic (in terms of hardware) solutions to the problem, but does this make them best? If we set m to 1, then any burst error can be accommodated to within three bits, but there may well be advantages in using larger m. Because of the interaction of the syndromes due to large errors, in some cases it appears that there are several solutions. Soft decisions, or cross-referencing with outer codes (in a block-interleaved message), may help us make the correct choice, but by using large symbols we can reduce drastically the probability of multiple solutions appearing. In effect, this means that decoding hardware (or software) can be simpler. Take, for example, the following sequence of syndromes in S_1 for $k = 7$:

$$00617325165430600617325...$$

The presence of the double zero ensures that we have a good starting point for finding the errors (assuming that the capacity of the code is not exceeded). Two zeros in a row suggest that at least one r' in the I_w window of syndromes is correct. Starting at this suspected good point we get the error sequence,

$$6, 6 + 1 = 7, 7 + 3 = 4, 4 + 2 = 6, 6 + 5 = 3, 3 + 1 = 2,$$
$$2 + 6 = 4, 4 + 5 = 1, 1 + 4 = 5, 5 + 3 = 6, 6 + 0 = 6$$

or
$$\xi_r = 6, 7, 4, 6, 3, 2, 4, 1, 5, 6, 6$$

Notice that, in the syndromes, there is also a single zero. This suggests that two adjacent errors were the same (6 and 6). If one more symbol in r' is corrupted, then we no longer have a double zero to guide us to a starting point. We may not be able to distinguish the starting zero from a double error of the same value which also produces a zero. In this case, we must perform a second check. The errors above give rise to the syndrome pairs

$$66, 77, 44, 66, 33, 22, 44, 11, 55, 66, \underline{66}$$

each overlapping by one symbol. The bold symbol represents the error of some $k + xI_w$ symbols previous in r' (where x is an integer) while the second symbol of the pair just tags along. If, however, we have started in the right place in a sequence of syndromes, then the last error we calculate (underlined) should be equal to the last syndrome (strike-through). If we add one further error to r' then the syndrome S_1 becomes

$$20617325165430420617325...$$

We no longer have a double zero to guide our starting point, but we do have two single zeros. Using the previous starting point we end up with a final error $6 + 4 = \mathbf{2}$ which also agrees with the next (last) syndrome. If we start at the other zero in the sequence we get the errors

$$4, 6, 6, 0, 1, 6, 5, 7, 2, 3, 5, 1$$

Now the final error is 1, but the final syndrome of this sequence is 3. Clearly there is a mistake and this cannot represent a valid solution. In this case, there are three bits per symbol, or eight possible symbols. If we take a simple approach and say that the incorrect solution will yield a random final error, then one in eight such random results will equal the last syndrome of the sequence and be indistinguishable from the correct solution. Using binary symbols ($m = 1$) then for 50% of the time we will be uncertain as to whether or not we have selected the correct starting position in the sequence. This is a rather simplistic view of the problem but by using larger symbols we can more easily extract such error information from the syndromes. Not only this, but the larger the symbols the smaller the probability of adjacent errors being equal and hence giving a zero in the syndrome in the first place.

It was previously stated that double zeros in S_1 give us a good starting place for calculating the errors from the syndromes. While it is a good starting position, it is not necessarily correct since triple errors of the same pattern will produce a double zero in the syndrome. Take for example the error in r', 3, 3, 3. This will result in the cyclic syndrome

$$... 0, 3, 0, 0, 3, 0, ...$$

This condition is easily detected since the isolated non-zero syndrome suggests an error in d', but this will not be matched by an impulse in S_0. While this is detectable, it means more complexity. By increasing m, we can reduce the probability of this scenario, but for binary symbols it is a very likely situation. All this points to the use of modest sized symbols to keep decoding complexity low.

We do not need a zero in S_1 to guide our decoding process, however. Provided there is no actual overlap of the syndromes at the beginning and end of a sequence (i.e. the capacity of the code isn't totally exceeded) we can use the rule above (i.e. last non-zero syndrome equals last error), to try all I_w solutions and see which pass the test. Using large symbols, we stand a good chance of only ending up with the correct solution, but in the event of multiple choices we can look to soft decisions or data context to help us. Since burst error capability costs mainly latency (rather than coding rate) we could avoid these tricky situations by using a larger k than is strictly necessary.

This scheme has been limited thus far to a coding rate approaching ½, but there is no reason why we can't reduce it. The fundamental requirement of error coding is to make valid messages appear as different as possible. A second transform can be applied to the data to generate a second redundant symbol stream. Such an encoder might appear as follows:

$$s_i = \sum_{j=-k}^{k} \alpha^{-j} d_{i+j}$$

The overall effect of errors in d' and s' on the two syndrome streams which result from this code is similar to the previous case. If the syndrome streams are called P_0 and P_1, an error in d' causes an impulse of k non-zero syndromes in P_0, and an isolated non-zero syndrome in P_1. The impulse is no longer composed of constant values equal to the error, but forms a ramp of increasing powers of α, with the central syndrome equal to the error. An error in s' causes a single syndrome in P_0 and a cyclic error pair in P_1, starting k symbols after the error. For an error ξ, the first symbol of the pair is $\xi\alpha^k$, while the second symbol is $\xi\alpha^{k+1}$. After each cycle of I_w syndromes these increase by α^{lw}. If the field size equals $2k + 1$ then the values remain unchanged each cycle, increasing by α^0. We can now correlate solutions from both redundant streams to yield the most probable result.

By adding a second interleaved redundancy stream another interesting possibility arises. If this stream is displaced from the other two, then in the event of a burst error, while d' and r' may be damaged, as will s' for some other part of the message, the message may still be recoverable from s' when it arrives a little later. To avoid adding more latency, the facility need only be used when gross errors occur, relying on d' and r' in normal use. These latter ideas are just to give you a little more food for thought. Whether or not they could be implemented really depends on the system constraints, but we do nonetheless end up with a very powerful yet simple code.

The programmable nature of this code means that we could quite easily create an adaptive coding system. In the mobile phone example we want small latency, and where error correction indicates few errors k can be kept small. If, however, the error rate starts increasing, k can be increased. The theory here is that at times, slightly delayed communication is better than broken communication or no communication at all. To implement such a scheme we still have to consider how to modify k without interrupting the decoding process.

Part Six

Practical considerations

13

Manipulating elements with hardware

13.1 MULTIPLYING ELEMENTS USING XOR GATES

You have already had some exposure to hardware appropriate for handling finite field elements. This is particularly true in the case of the CRC generator/checker circuit. By making a few modifications to the basic CRC checker such as removing the input, then for $x^4 + x^3 + 1 = 0$ we have the circuit of Fig. 13.1. If this circuit is preloaded with a non-zero value then successive clocks will repeatedly cycle it through the sequence of field elements. By preloading with 2, we generate the sequence

$$2, 4, 8, 9, 11, 15, 7, 14, 5, 10, 13, 3, 6, 12, 1, 2, 4, \dots$$

which is consistent with increasing powers of α for this field.

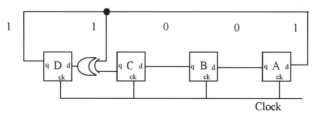

Fig. 13.1 Generating a finite field using hardware.

The reason that this circuit behaves in the way it does is because the XOR gate actually multiplies the current register contents by α. We can extract this part of the circuit as shown in Fig. 13.2 to make a multiplier which takes a finite field element in the form of a pattern and returns the result in the form of a pattern. As far as manipulating elements is concerned a choice must be made between representing the elements as their binary values (as in this case) or as their powers; $\alpha = 2$, could be represented by 1 which is the power, or 2 which is the binary value. When multiplying elements representation by powers can be convenient since multiplication and division is simply addition and subtraction of the powers. When adding values, however, we have to XOR the binary representations. Certainly when tackling experiments in software a hybrid representation proves most convenient, moving between representations via look-up tables appropriate to the function required. When using powers caution is required though since there are no adequate means of expressing the 0 element of the field.

In the Pascal libraries given, the function AxB (a, b) expects values as powers but nonetheless treats the power zero as the zero element of the field, and will return zero if either a or b is zero. While this may seem anomalous, it is a programming convenience. For implementing the FFT increasing powers of α are required, and the function ApB (a, b) is used. In this latter case, a and b are again powers, returning a value of α appropriate to raising a to the power of b, but if $b = 0$ then the result will always be 1. Treatment of the zero element can, therefore, cause confusion if due consideration is not given to the problem.

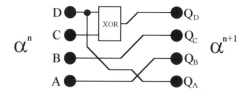

Fig. 13.2 Dedicated multiplier for $GF(2^4)$, $x^4 + x^3 + 1 = 0$.

The circuit of Fig. 13.2 takes a pattern representation of a field element and multiplies it by α. To multiply by α^2 we could place two of these circuits in series, but by the time we got to a multiplier for α^{14} the circuit would be needlessly large and slow. Figure 13.1 shows us how we can implement a multiply serially by preloading our value into the registers and clocking once for each power of α that is required. Again this is slow but a little more economic on hardware. Both these approaches demonstrate how we can easily generate the whole set of 14 circuits in a faster and more economic form. In Table 13.1 each row defines how to calculate the four multiplier outputs, Q_d to Q_a, based on the four inputs A, B, C and D. The multiplication factor is determined by the column. In any column the inputs are found by applying the circuit in Fig. 13.2 to the previous (left-hand) column. Where several terms occupy a box, they must be summed or eXclusive ORed together to find their combined weight, whether it is even (0) or odd (1). As ever, similar terms cancel out when added.

Table 13.1 Calculating the multiplier circuits over GF(2^4)

	α^1	α^2	α^3	α^4	α^5	α^6	α^7	α^8
$Q_d =$	CD	BCD	ABCD	ABC	ABD	AC	BD	ACD
$Q_c =$	B	A	D	CD	BCD	ABCD	ABC	ABD
$Q_b =$	A	D	CD	BCD	ABCD	ABC	ABD	AC
$Q_a =$	D	CD	BCD	ABCD	ABC	ABD	AC	BD

	α^9	α^{10}	α^{11}	α^{12}	α^{13}	α^{14}	$\alpha^{15}(1)$	α^1
$Q_d =$	BC	AB	AD	C	B	A	D	CD
$Q_c =$	AC	BD	ACD	BC	AB	AD	C	B
$Q_b =$	BD	ACD	BC	AB	AD	C	B	A
$Q_a =$	ACD	BC	AB	AD	C	B	A	D

Taking α^{13} as an example, we would end up with the following circuit in Fig. 13.3.

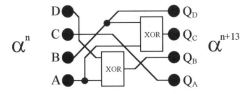

Fig. 13.3 Constructing a multiplier from Table 13.1.

Dividers use exactly the same circuits. Dividing by α for example is the same as multiplying by α^{14}, or more generally, dividing by α^n is the same as multiplying by α^{15-n}. Sometimes we require programmable multipliers and dividers. These can be achieved in several ways, two of which are illustrated in Fig. 13.4.

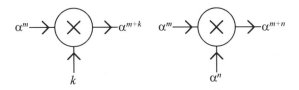

Fig. 13.4 Programmable multiplier configurations.

To implement the first of these we need to construct a set of equations defining the relationship between k (the multiplier power) and α. Table 13.1 shows us how to begin this process. If we examine the contribution of the pattern inputs (α^m) $ABCD$ to Q_a for example, with respect to k then we have the following in Table 13.2.

Table 13.2 Calculation of Q_a from k and $ABCD$ using Table 13.1

k	0	1	2	3	4	5	6	7	8	9	10	11	12	13	14	15
k_0	0	1	0	1	0	1	0	1	0	1	0	1	0	1	0	1
k_1	0	0	1	1	0	0	1	1	0	0	1	1	0	0	1	1
k_2	0	0	0	0	1	1	1	1	0	0	0	0	1	1	1	1
k_3	0	0	0	0	0	0	0	0	1	1	1	1	1	1	1	1
A	✓	✗	✗	✗	✓	✓	✓	✓	✗	✓	✗	✓	✓	✗	✗	✓
B	✗	✗	✗	✓	✓	✓	✓	✗	✓	✗	✓	✓	✗	✗	✓	✗
C	✗	✗	✓	✓	✓	✓	✗	✓	✗	✓	✓	✗	✗	✓	✗	✗
D	✗	✓	✓	✓	✓	✗	✓	✗	✓	✓	✗	✗	✓	✗	✗	✗

In Table 13.2 a tick indicates that for a particular value of k, the input (A, B, C or D) is used in the creation of the Q_a output. From this we can derive a Boolean expression linking k, A, B, C and D to Q_a using the Karnaugh maps of Fig. 13.5. This gives rise to (13.1).

A	k_1k_0	01	11	10
k_3k_2	1	0	0	0
01	1	1	1	1
11	1	0	1	0
10	0	1	1	0

B	k_1k_0	01	11	10
k_3k_2	0	0	1	0
01	1	1	0	1
11	0	0	0	1
10	1	0	1	1

C	k_1k_0	01	11	10
k_3k_2	0	0	1	1
01	1	1	1	0
11	0	1	0	0
10	0	1	0	1

D	k_1k_0	01	11	10
k_3k_2	0	1	1	1
01	1	0	0	1
11	1	0	0	0
10	1	1	0	0

Fig. 13.5 Generation of Q_a from A, B, C and D.

In (13.1) it is important not to confuse + with the XOR operation used on symbols. Here + represents logical ORing while $\oplus$ represents the XOR operation. Clearly we have to generate four such expressions, one for each bit of the output Q. Careful observation of Table 13.1 reveals that many of the inner sum of product terms are shared, necessitating the generation of only seven to complete this example rather than the expected 16. For example, the contribution of input A to Q_a is identical to the contribution of input B to Q_b.

$$
\begin{aligned}
Q_a = A&\left(\bar{k}_3 k_2 + \bar{k}_3 \bar{k}_1 \bar{k}_0 + k_2 \bar{k}_1 \bar{k}_0 + k_3 k_1 k_0\right) \\
\oplus B&\left(\bar{k}_3 k_2 \bar{k}_1 + k_3 \bar{k}_2 \bar{k}_0 + \bar{k}_2 k_1 k_0 + k_3 k_1 \bar{k}_0 + k_2 k_1 \bar{k}_0\right) \\
\oplus C&\left(\bar{k}_3 k_2 \bar{k}_1 + k_3 \bar{k}_1 k_0 + \bar{k}_3 k_1 k_0 + \bar{k}_2 k_1 k_0\right) \\
\oplus D&\left(k_3 \bar{k}_1 k_0 + \bar{k}_3 k_2 \bar{k}_0 + \bar{k}_3 \bar{k}_2 k_1 + \bar{k}_2 \bar{k}_1 k_0\right) \qquad (13.1)
\end{aligned}
$$

Where we wish to multiply two unknown symbols that are both expressed as patterns, then we must turn to the primitive polynomial which describes the field, for help. Suppose our two inputs are α^m and α^n where the bits for each symbol are m_{0-3} and n_{0-3}. First we must AND all combinations of two bits (one from each symbol), sorting them into groups of similar powers. Consider Table 13.3 below.

Table 13.3 Generating a multiplier using the primitive polynomial

Power	Contributing terms	Equivalent powers	Additional terms
x^0	$m_0.n_0$		$m_1.n_3, m_2.n_2, m_3.n_1,$ $m_2.n_3, m_3.n_2, m_3.n_3$
x^1	$m_0.n_1, m_1.n_0$		$m_2.n_3, m_3.n_2, m_3.n_3$
x^2	$m_0.n_2, m_1.n_1, m_2.n_0$		$m_3.n_3$
x^3	$m_0.n_3, m_1.n_2, m_2.n_1,$ $m_3.n_0$		$m_1.n_3, m_2.n_2, m_3.n_1,$ $m_2.n_3, m_3.n_2, m_3.n_3$
x^4	$m_1.n_3, m_2.n_2, m_3.n_1$	$\equiv x^3 + 1$	
x^5	$m_2.n_3, m_3.n_2$	$\equiv x^3 + x + 1$	
x^6	$m_3.n_3$	$\equiv x^3 + x^2 + x + 1$	

Clearly if we multiply the least significant bits of each input word, m_0 and n_0, then we end up with a result of significance x^0, or the least significant bit in the result. This is fine for powers up to x^3 in this case, but with a four-bit result expected, we have to fold higher powers back down using the primitive polynomial for the field. From $x^4 + x^3 + 1 = 0$, we arrive at the equivalent powers column of Table 13.3. Where these equivalent terms emerge, they must be added to the appropriate power row. In all cases the equivalent powers contain x^3 and 1, so all terms in the contributing terms column must be added to the x^0 and x^3 rows. x^5 contains the additional term x, so these terms also appear in the x^1 row and so forth. Where high terms are redistributed back in to the lower terms by the polynomial they are placed in the additional terms column to the right of the table.

To compute the multiplied output (x^0 to x^3), all the bit pairs in rows x^0 to x^3 are ANDed, and the results XORed. The final circuit is shown in Fig. 13.6.

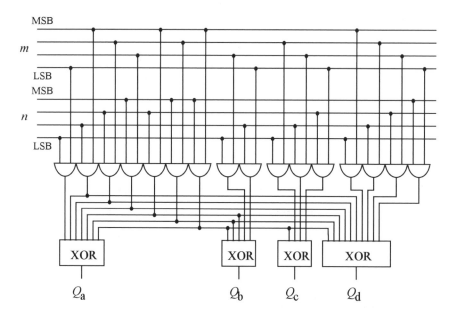

Fig. 13.6 Constructing a general element pattern multiplier.

Over large fields this approach could become cumbersome except in dedicated applications such as a high-speed Fourier transformer where full custom IC design is available. Incidentally, executing a forward Fourier transform three times on a set of data actually has the effect of performing an inverse Fourier transform. Where speed is not an issue this can save a considerable amount of circuitry especially in parallel implementations. Alternatively an inverse Fourier transform can be performed directly by a Fourier transform if the output symbols are reversed in significance such that $d_0 = F_0$, $d_{14} = F_1$, $d_{13} = F_2$ etc. (over GF(2^4)).

13.2 A HARDWARE TIME DOMAIN ENCODER/DECODER

Figure 13.7 shows how just a few simple elements are necessary for single-symbol error correction. If we consider a single-symbol error correction over GF(2^4) in the time domain then we have the following requirements for initial encoding:

We must calculate two check symbols R and S such that

$$\sum_{i=1}^{15} d_i = 0 \quad \text{and} \quad \sum_{i=1}^{15} d_i \alpha^i = 0$$

where R and S replace d_{14} and d_{15}.

If we initially set R and S to 0, and evaluate two syndromes S_0 and S_1, much as we would when correcting errors, then

$$S_0 = R + S \quad \text{and} \quad S_1 = \alpha^{14} R + S$$

Rearranging gives

$$R = (S_0 + S_1)/(1 + \alpha^{14})$$

$$R = (S_0 + S_1)/\alpha^{11} \quad \text{(using the polynomial 11001)}$$

and

$$S = S_0 + R$$

The data are input in reverse order (d_{13} to d_1), and the register/XOR gate combination evaluates the two syndromes, S_0 and S_1, by feeding the register outputs back to their inputs. The registers must be reset before the operation starts and subsequently clocked once per input symbol. The output of the lower four registers is multiplied by α before feeding back. The two results are summed in four XOR gates and the result is multiplied by α^4 (from Table 13.1) which effects the divide by α^{11}. This produces R which is added to S_0 to give S.

Decoding uses a very similar circuit to Fig. 13.7 although its operation is slightly different. In this case all 15 symbols are clocked in (starting with d_{15} or S) having first reset the eight registers. This means that much of the circuit can be shared between encoding and decoding. If the two syndromes (the contents of the registers) are not zero after the last symbol (d_1) has been clocked in, then correction must be applied. Life gets a little tricky here because we have to perform a variable divide to evaluate the locator $k = S_1/S_0$ depending on S_0. If we rearrange the expression to $S_0 k = S_1$ then, remarkably, little further hardware is needed. The syndrome S_1 is located in the lower four registers of Fig. 13.7, having been derived from

$$\sum_{i=1}^{15} d_i \alpha^i = S_1$$

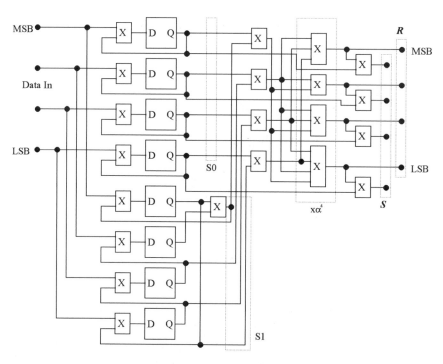

Fig. 13.7 Hardware check symbol calculator for *R* and *S*.

The lower four registers have a multiplier already connected to them such that each time they are clocked their output will be α times its previous value. To calculate k, therefore, the four data inputs must be set to zero following d_1 and the lower four registers clocked until their outputs equal the content of the upper four registers (S_0). The number of clocks required is therefore $15 - k$. The circuit of Fig. 13.7 is broken down schematically into Fig. 13.8 to show the relationships between the various parts more clearly. A 15 symbol data buffer (FIFO) has been added to the circuit so that during decoding the data can be corrected.

For decoding, the sequence of operations starts by strobing the clear input, then the 15 data plus check symbols are input in the order d_{15} to d_1, strobing the encode/decode clock once per symbol. When the final symbol has been clocked in the entire message will be present in the shift register at the bottom. The data inputs are set to zero and the encode/decode clock is clocked a further 15 times. If no errors have occurred, both S_0 and S_1 will be zero. The sum of the two syndromes will also be zero resulting in the output multiplexer being switched to the sum of the data plus S_0. Since this syndrome S_0 is zero this will have no effect.

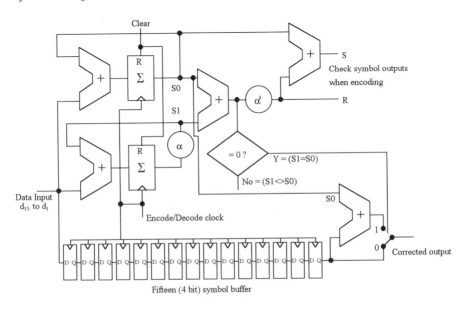

Fig. 13.8 Schematic of the encoder/decoder circuit.

If an error has occurred then the syndromes will not be zero. In this instance the encode/decode clock is again clocked 15 times and for one of these periods the syndromes will become equal as S_1 cycles round the 15 non-zero field elements. At this point S_0 will be summed with the current output of the buffer to correct it. For the other 14 clock periods the multiplexer will select the direct buffer output. This gives rise to a very neat solution capable of sharing much of the encode and decode circuitry.

Since for large data blocks the data will inevitably arrive in a serial manner, serial solutions to encoding and decoding will often be as fast as parallel solutions since the result cannot be known until the last data element arrives in either case. Had the calculation performed by the circuit of Fig. 13.7 been done in a parallel manner, generating R and S from weighted sums of the 13 input symbols, the results would be available no sooner but much more hardware would be required. This is true even in the case of frequency domain error coding where a full parallel implementation of a Fourier transformer might appear advantageous. As the data arrive, a partially parallel implementation of the Fourier transformer, shown in Fig. 13.9, will allow the FFT result to be available immediately after the arrival of the last symbol. In the event of an error recursive extension must be used to regenerate the completed error spectrum, so again, the symbols arrive serially and nothing can be done to speed up the subsequent inverse Fourier transform.

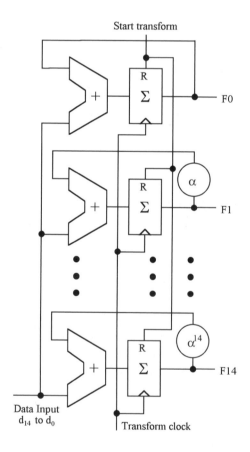

Fig. 13.9 Partially parallel Fourier transformer for GF(2⁴).

If mixed domain encoding is used then the corrected result is now available. If a completely frequency domain solution is proposed then at this point alone will a fully parallel FFT circuit be of any speed advantage, and at massive cost. As soon as the last recursively extended symbol appears, the completed error pattern becomes available from the inverse Fourier transformer. If the data are accessible in a parallel format then parallel correction is possible, and the final FFT could conceivably be done in parallel.

In short, for a frequency domain encoder/decoder one out of four Fourier transforms is amenable to a fully parallel implementation. For mixed time/frequency encoding neither of the two potential Fourier transforms can usefully benefit from being done in parallel.

In Fig. 13.9 data arrive in reverse order and each of the multiply accumulate circuits evaluates one of the 15 frequency components. The inverse Fourier transform can use a largely identical circuit. The only difference is that the outputs appear in different places. F_0 remains the same, but F_{14} and F_1 are swapped as are F_{13} and F_2 and so on. Where the FFT output is required serially, a sideways loadable shift register will be needed, and with a very modest amount of multiplexing, the majority of the circuit can be shared between the forward and inverse transform. An alternative strategy might be to use exactly the same transformer hardware, but reverse the order in which the data are input. Using the circuit of Fig. 13.9, the forward Fourier transform is performed by inputting the data in the order d_{14} to d_0, whereas the inverse Fourier transform effected by inputting the data as F_0 to F_{14}.

13.3 GENERAL HARDWARE CONSIDERATIONS

Almost all aspects of the error control mechanism that is adopted will have a major bearing upon the hardware that will be required for its implementation. The field size probably requires the most consideration since this permeates every aspect of the hardware. It should be obvious enough now that everything scales in some form or other with the field size. If we operate over $GF(2^m)$ and $q = 2^m - 1$ then the element bit widths scale linearly with m while functions like the Fourier transform scale with q^2 where q may be in time or size or both (as in Fig. 13.9). For this reason some space has been devoted previously to packaging strategies which increase the message size for a given q (multi-dimensional data structures) and increase the error capabilities of limited codes (block-interleaving) and coding methods which decrease the number of transforms required (mixed time/frequency domain coding). We are fortunate in that most of the calculations that are required can be implemented in parallel, mixed parallel/serial and fully serial formats. This gives us a great deal of choice and the ability to trade-off factors like speed, cost, size and flexibility.

The technology upon which the proposed solution is built will be the deciding factor in many of the equations that have to be balanced. Where micro-controller cores and memory are available the solution can be the ideal balance between software for the mundane tasks and hardware accelerators for the parallel operations. Using more modest technology, external memories may be required and the processing less flexible. Even so, the previous example shows that very little hardware is required in some cases. At the opposite end of the spectrum the solution may be completely software. In this case time probably becomes the dominant issue, setting the limits of our scope. Using the Fourier transformer of Fig. 13.9 as an example, Table 13.4 offers an approximate guide to the hardware resources that will be required for various field sizes. Because resources like multipliers are dependent upon the polynomial as well as the field size, a guess has to be made.

Table 13.4   Hardware resources for a parallel/serial Fourier transformer over GF(2^m)

m	One-bit regs	Two-bit XOR	Clocks	Message (bits)	Cost (a)	Cost (b)	Cost (c)
3	21	~36	7	21	2.71	0.33	8.98
4	60	~128	15	60	3.13	0.25	7.88
5	186	~431	31	186	3.32	0.17	6.55
6	378	~1152	63	378	4.05	0.17	7.28
7	889	~3136	127	889	4.52	0.14	7.18
8	2040	~8192	255	2040	5.02	0.13	7.49

Since a Fourier transformed message is the same size as the original, the circuit must have the capacity to store the entire message. This is evident in the one-bit registers column which is equal to the message size. Many of the XOR functions will be more than two bits so the operation has been broken down into an equivalent circuit using two-input devices. The clocks column is a measure of execution time. Since the circuit is partially serial the execution time is the same as the time taken to serially clock the message symbols (each *m* bits) into the circuit. Because the capacities of many VLSI technologies are measured in terms of gates and registers, a table such as 13.4 will be a good guide as to what can and cannot be fitted in.

Three measures of cost are included in the table. Cost (a) attempts to define some form of hardware efficiency. The registers and XOR gates are added (giving them an equal weighting) and the result divided by the number of message bits. The units of this cost are therefore something like registers per **bit** of work done. Cost (b) is a measure of the time or number of processing cycles required per bit. The final cost represents a combination of hardware and time per bit of processing. Cost (b) was multiplied by 19 (to bring the averages of costs (a) and (b) together), and the two costs summed. Cost (c) therefore assumes equal importance is placed on hardware and processing time. While these measures are a little arbitrary, they serve to illustrate at least one possible way of measuring some of the system requirements.

A fully serial Fourier transformer will increase mostly with bit size *m* rather than with field size, but the cost of this is an increase in execution time. The schematic of a fully serial Fourier transformer is given in Fig. 13.10.

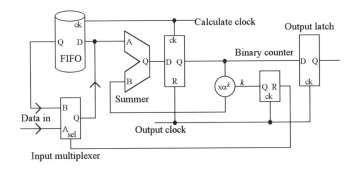

Fig. 13.10 A fully serial Fourier transformer.

Data are input to the circuit in reverse order, and during this time the input multiplexer is set to A and the FIFO is loaded with the data. A binary counter generates the frequency component index and is reset during this input phase. The calculate clock is clocked once for each input symbol. When all symbols have been input the multiplexer is switched from the data input to the FIFO which repeats the input symbol sequence for the calculation of each frequency element. When all data symbols have been input, F_0 will appear on the output of the summer latch. This is now clocked into the output latch and the counter incremented. Using the FIFO, the data are cycled through the circuit again for the evaluation of the next frequency element. For each frequency element the index increases, and so does the feedback multiplier in the summation path. Table 13.5 indicates the hardware resources that this circuit will require.

Table 13.5 Hardware requirements for a serial Fourier transformer over GF(2^m)

m	One-bit regs	Two-bit XOR	NAND	Clocks	FIFO (bits)	Cost (a)	Cost (b)	Cost (c)
3	9	~12	21	49	21	2.00	2.33	11.10
4	12	~22	32	225	60	1.10	3.75	6.87
5	15	~36	45	961	155	0.62	6.20	4.22
6	18	~51	60	3969	378	0.34	10.50	5.02
7	21	~72	77	16129	889	0.19	18.14	6.61
8	24	~150	96	65025	2040	0.13	31.88	10.55

As with Table 13.4 some figures are approximate since they depend upon the polynomial used in the generation of the field. Also the assumption that all hardware elements count equally has been made. In reality this won't be entirely true since a NAND gate uses about one-quarter of the resources needed by an XOR gate or a register. The columns entitled cost have been calculated similarly to those of Table 13.4. The final cost column (c) has been weighted so as to be comparable to that of Table 13.4. It is interesting that in both examples, a field size of GF(2^5) gives the least cost per bit. The cost of the FIFO has been excluded from the calculations in this example. The reason for this is that on a low-cost technology it may well have to be external and on a standard cell-based design it will almost certainly be precompiled and highly optimized. Typically it is the random logic functions that use up most of the resources.

13.4 BIT ERROR RATES

A practical consideration that has so far received little attention is the correcting power of the coding required to do the job. In other words, if we need a certain improvement in channel performance, how many bits must we be able to correct? This aspect of the subject gets quite mathematical, so I shall give only a simplified view of the problem in keeping with my overall aim of minimizing the maths. The quality of a channel is often defined in terms of its bit error rate, or BER. This is a measure of how likely a bit is to be in error. A digital mobile telephone inside a building will experience a raw BER of about 10^{-2}. This simply means that on average, one bit in 100 will be in error, or the probability that any bit will be incorrect is 0.01.

Suppose that we have such a channel where p, the probability of a bit error, is 0.01. Let's examine a five-bit message where the bits are denoted a, b, c, d and e. The probability that bit a only, is in error is given by

$$P(a) = p \times (1-p) \times (1-p) \times (1-p) \times (1-p)$$

i.e. the probability that a is in error and b, c, d and e are not. The probability of bits b, c, d or e only, being in error is the same. The overall probability that the message contains a one-bit error is, therefore, the summation of the probabilities that any single bit is in error. In this example, the probability of the message containing a single-bit error is

$$P(1) = 5 \times p \times (1-p)^4$$

so if p is 0.01, this works out to about 0.048 03. So what is the probability of a two-bit error? The probability of bits a and b being in error is given by

$$P(a, b) = p \times p \times (1-p) \times (1-p) \times (1-p)$$

This works out to 0.000 097 03. So how many combinations of two bits are there in a five-bit message? By hand, we can find a/b, a/c, a/d, a/e, b/c, b/d, b/e, c/d, c/e and d/e, or 10. Fortunately we can draw on the following equation to determine this:

$$\text{Combinations of } m \text{ bits in } n \text{ total} = \frac{n!}{m!(n-m)!}$$

In our case, $n=5$ and $m=2$ so we get $120/(2 \times 6) = 10$. Generalizing the above ideas we end up with a probability of m bit errors in an n bit message being given by

$$P(m,n) = p^m (1-p)^{n-m} \frac{n!}{m!(n-m)!}$$

In our example,

$$P(0, 5) = 0.951$$
$$P(1, 5) = 0.048\ 03$$
$$P(2, 5) = 0.000\ 970\ 3$$
$$P(3, 5) = 0.000\ 009\ 801$$
$$P(4, 5) = 0.000\ 000\ 049\ 5$$
$$P(5, 5) = 0.000\ 000\ 0001$$

If we add all these probabilities, we should get 1, and give or take a few rounding errors this is so. The probability that the message arrives error-free is 0.951, and the probability that it arrives in error is $(1 - 0.951)$ or 0.049. This is roughly one in 20.5 messages. We could have predicted that with a BER of one bit in 100, one out of 20 five-bit messages would contain an error. As there is a finite probability that two or more error bits may be contained within a single five-bit message, the actual number of messages corrupted is slightly less than one in 20.

We now need to establish how this information can be used to work out how much error correction capability will be required. Suppose we are working over $GF(2^6)$ and our message block is 378 bits. The probabilities of bit errors in the block (with $p = 0.01$) are listed in Table 13.6 below.

Table 13.6 Error probability in message block

Bit errors	Block error	Sum	Blocks
0	0.0224	0.0224	1.02
1	0.0855	0.1079	1.12
2	0.1628	0.2707	1.37
3	0.2061	0.4768	1.19
4	0.1952	0.6720	3.05
5	0.1475	0.8195	5.54
6	0.0926	0.9121	11.38
7	0.0497	0.9618	26.18
8	0.0200	0.9818	54.95
9	0.0096	0.9914	116.28
10	0.0036	0.9950	200.00

The table lists in the *Block error* column, the probability that *Bit errors* will have occurred in the message. The third column, *Sum*, is a summation of the *Block error* column, while the last column, *Blocks*, is found from the reciprocal of one minus *Sum*. The *Blocks* column tells us how many messages will contain *Bit errors*. In other words, if we look at the last row, this tells us that only one message in 200 will contain more than ten bits in error.

Suppose that we want to get the BER down to one bit in 10 000 from one in 100 by using error coding. Ten thousand bits span about 26.5 message blocks. From Table 13.6, you can see that about 54 out of 55 messages will contain eight or fewer bit errors, so if we set up an eight-symbol correcting code (assuming a worst-case single-bit error per symbol) then we will ensure that 54 out of 55 messages will be completely error-free after correction. If we went for seven bits (25 out of 26 messages) we would be just outside our target.

I have made the gross approximation that an unrecoverable message counts as one bit error. In fact it will represent a sudden error burst in the channel equivalent to the number of error bits in the message. Suppose that the one in 55 messages that could not be recovered contained nine error bits. In this case we would have a BER of $9/(55 \times 378)$ or one bit in 2310 on average (assuming our attempts at correction did not make matters worse). At first, this seems poor at the side of the intended one bit in 10 000, but many such systems are augmented by an automatic repeat request facility. In the event of an unrecoverable error, the message can be retransmitted. Not only this, but there is a finite probability (that you may wish to determine) that some of the bit errors will occur within a single symbol. Don't forget that with an eight-symbol correcting code over $GF(2^6)$ we could correct up to 48 bits per message best case. Given these facts, the real BER may be much lower than we expect, stemming only from occasional miscorrection of very large errors.

In this section, the calculations have been based upon random errors rather than burst errors. Where burst errors are of interest a good knowledge of their cause is essential if protection levels are to be determined. As such, little advice can be given here, although an example in Chapter 14 may provide some insight. In the digital mobile phone example above, many errors are likely to be bursty in nature so the proposed encoding may work much better than predicted here.

14

Applications

So far you will have discovered how to use some of the more common tools available for error control. Which tool you use in any situation is up to you, and only careful consideration of the application and practice can help you make the choice. In this chapter, two very loosely defined applications will be examined, and we'll consider the reasons why one solution might be better than another. This is the real 'engineering' part of the subject where ideas must be put into use.

Where a problem is very specific the most appropriate solution is usually fairly obvious. For example, if a large data block requires only minimal error protection but the process must be quick, then a multi-dimensional single-error correcting code over $GF(2^4)$ would probably be the best solution. No costly (both in time and hardware) transformations are required for encoding or correcting, and by using a small field a hard-wired solution is straightforward. A more general solution with high or variable levels of protection inevitably leads one towards a frequency domain-based approach with the need for an integral micro-controller to manage the processing.

A knowledge of the frequency and types of errors that are likely to occur, and their source, can also be invaluable. Where large losses are anticipated then the coding must be augmented by packaging strategies like block-interleaving and perhaps the addition of inner codes. If the errors are small but spread out, then a small field with a high symbol-correcting capability will be best. In this case block-interleaving won't help at all.

Latency is also another factor that must be considered. This is a measure of the time it takes to code or decode a block of data. While the average throughput of data may be high, it can still take a long time between receiving the coded data and providing an output. This is a particular problem with block codes since the first bit of data in a block to arrive cannot be output until the last bit of the block has been received, and the whole subsequently decoded. Convolutional coding wins here since there is no requirement to construct blocks. These same factors also determine the memory that will be required by the processing. If a dedicated, single-chip solution is required then the available on-chip memory has a direct bearing on the possible solutions.

An interesting example of the latency problem is nicely encapsulated in the proposed digital audio broadcasting (DAB) standard. Figures of around one second have been suggested as necessary in order to reconstruct the audio signal from the transmission. While for most of the time this will not make much difference, it will offset the hourly time signal. To pre-emphasize the time signal by one second assumes a constant decoding latency at the receiver which is unlikely to be the case if multiple sources for the chip sets appear.

14.1 EXAMPLE APPLICATION ONE

The first test case has the following specifications:

> Must be high speed
> Low cost
> Protect 8 Kbit blocks of data with small overheads
> Very few errors likely

Often the key requirements of a system will have some sort of priority. Cost might be the overriding factor, or speed might be. In **real-time** systems, guaranteed speed in all circumstances will be paramount. In a mass-produced disk drive or network card, cost will be a much more sensitive issue. Generally, 'very few errors' would be more accurately defined as a probability of errors, or a **bit error rate** (BER). The specification would define the existing channel characteristic and demand an improvement of so many decibels. In this example we'll assume that this means that we only need a single-symbol correction capability over each encoded block.

The fundamental question will be 'block code or convolutional code?' The requirement for small overheads or redundancy makes block codes the most viable option since these codes can be scaled to large data blocks with little extra overheads. This choice is fundamental because it affects every other aspect of the solution as the codes are so different in operation.

Having made this choice, the next question will be 'time domain (TD) or frequency domain (FD)?' In this example almost all system specifications will answer this question. First, the system must be high speed. FD coding involves very large processing overheads, increasing by roughly a power of two with the field size. Complexity is involved in both the encoding and decoding of data. To make this process high speed would require power-hungry dedicated hardware, and lots of it. TD processing scales roughly linearly with block size.

Second, the system must be low cost. Ideally this means we want to be able to implement it on a low-cost programmable device. FD coding requires considerable memory for construction of things like the error spectrum and temporary storage of transformed data. Most low-cost programmable technologies cannot provide much memory which means moving to a more expensive technology or using support chippery. TD processing requires buffering the data block once only, in order to be able to correct it before outputting, plus minimal extra storage for the syndromes and check symbols.

The third specification has little bearing on the processing since either the time domain or the frequency domain can support these kinds of data block sizes with comparable levels of redundancy. The last factor, however, is the crux of the matter. Only few errors will occur. The whole point of FD coding is to make the solution of high or variable levels of error protection more straightforward. Neither high levels of protection, nor flexibility are required here making TD processing the obvious choice.

Now we must decide how to structure the data block. Will a one- or a multi-dimensional strategy be best? It often helps to remember why we have a choice. The point of multi-dimensional data structures is to allow the use of small fields and hence small numbers of elements, over large data blocks. Why is this useful? Because in dedicated hard-wired (high-speed) solutions we need only to support a small number of different processing elements. The number of elements increases linearly with field size, while their bit widths scale with the $\log_2$ of the field size. The number of binary ANDed terms in an element multiplier increases as the square of the number of bits in each element.

With this information we'll probably choose a multi-dimensional format but a few sums should be considered first. Speed may be an issue here. Forming the data into an n-dimensional structure involves calculating check symbols for each of the n dimensions, hence more work during encoding. Careful structuring of the problem (i.e. doing adds before multiplies etc.), however, means that the increase in calculations can be pushed into hardware, incurring no extra time overheads over single dimensional data.

In conjunction with the coding strategy above, we must consider block-interleaving. Bearing in mind the point of block-interleaving, decorrelating or distributing large errors between small independently encoded blocks of data, it is clear that there is no need for it here since we only expect small and infrequent (random) errors. Figure 14.1 shows the design process.

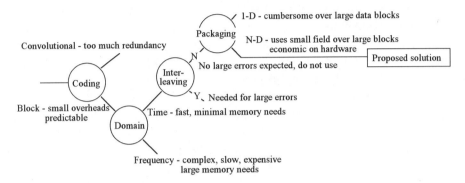

Fig. 14.1 The design process for application one.

Having decided how processing is to be done, we now need to look at the details. GF(2^4) uses only 16 symbols (including 0), and does not need to be extended into many dimensions in order to accommodate 8 Kbits of data. Since each symbol will be four bits, 8 K bits represents 2 K symbols. Each dimension in GF(2^4) can be no more than 15 symbols, so we'll need n dimensions where

$$15^n \geq 2K + n + 1$$

The $n + 1$ allows for the addition of check symbols (one per dimension plus one for error pattern). Solving, the smallest usable n is 3 giving a largest block size of 3375 symbols as opposed to the 2052 required. We do not need to worry about the extra capacity, we can simply shorten the code by assuming the data are zero in these locations so they do not enter into the time domain computations. The positioning of the empty data within the three-dimensional block structure is arbitrary and we can choose it for maximum hardware/processing convenience. We could use only 13 symbols per dimension, possibly economizing on multiplier circuits, or we might simply leave the last chunk of the final dimension empty.

So how close does this solution come to the specification? The first requirement was that the device should be high speed. Using the recommended approach, the check symbols can be made available only a few nanoseconds after the final bit has been input to the block, and similarly the syndromes can be found as quickly. There will be a latency of one block, however, since in the event of an error correction can't take place until the whole block is received. Convolutional coding would have had the edge on this point.

The second requirement was for low cost. The proposed implementation is certainly very simple and will easily fit on a very humble programmable gate technology. An external memory buffer may be required, but if the programmable device can support fractionally over 1 KByte of internal memory then even this is not necessary.

The third requirement was for small overheads. We have added four symbols to a 1 KByte message which is a redundancy of 0.2%! This is quite impressive. The requirement that we need cater only for a few errors made life simple. In this example we can tolerate up to four bits in error provided they are contained within a single symbol.

14.2 EXAMPLE APPLICATION TWO

In this second example the error coding problem is to protect blocks of data that are to be stored on an optical medium. While small surface defects are compensated for by the optical system, large defects are not and result in large losses of adjacent bits. Each data block is 320 bytes and the storage manufacturer would like to be able to recover from losses of up to 15 consecutive bytes of data worst-case, in a single block.

The natural block structure of this problem leads us immediately to block codes, and the large consecutive data loss problem suggests that block-interleaving will be required in order to keep the codes manageable. Since block-interleaving is to be used, the block will be split into smaller chunks in any case, so we may not have to resort to multi-dimensional processing.

Consider first bit-wise interleaving where the symbols are horizontal. Giving the block-interleaved block a square aspect ratio means about 51 bits (just under in fact) in each dimension. The nearest field that can support 51 bits in a single block is $GF(2^4)$, but we must also allow for redundancy. $GF(2^4)$ will cover a block of 60 bits which leaves only nine bits, enough for only a single symbol correcting code. If we squeeze the block inwards so that it is only 44 bits wide, then we will be able to have a double symbol correcting outer code, and the block will be 59 bits high. If the block is 59 bits high there is no room for an inner code over GF(2^4) but we may not need one. Our requirement is to be able to recover from 15 consecutive bytes of data loss which is 120 bits.

We can construct the block slightly larger than is necessary, 44 bits by 60 bits (330 bytes). If the block is 60 bits high (7.5 bytes) then a worst-case error of 15 bytes will at most spread across only two columns of bits in any row. At worst these two columns could be at a horizontal symbol boundary resulting in a two-symbol error and we can recover from this. In fact, all consecutive errors up to 37.5 bytes (five columns of bits) will be corrected. If the errors were exceedingly obliging and lined up exactly with the horizontal symbol boundaries, then an error of up to 60 consecutive bytes could be recovered, since still only two horizontal symbols would have been affected. The final block size is thus 3600 bits with a redundancy of about 29%, illustrated in Fig. 14.2.

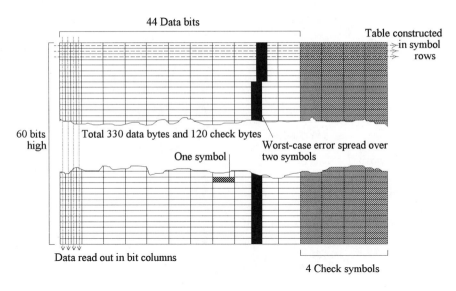

Fig. 14.2 Possible (bit-wise) data structure for application two.

Because the data area is 330 bytes rather than 320, the redundancy is slightly worse than might be expected. The extra ten bytes came from the requirement to ensure that 15 bytes of consecutive errors would spread only over two columns, so the columns had to be 60 bits high rather than 59.

If symbol-wise interleaving is used the block must be constructed slightly differently. In order to maintain a roughly square block, it is now more appropriate to operate over GF(2^5); 320 bytes gives us a block of about 23 symbols square. Since each row can contain 31 symbols, we might modify the block to 27 horizontal by 19 vertical symbols (allowing four check symbols per row). One column of symbols is thus 95 bits, or nearly 12 bytes. Interestingly, this is very near our target of 15 bytes, and in a single column; 15 bytes' worth of five-bit symbols is 24 symbols. If we construct the table 24 symbols deep and 22 symbols across (330 bytes), then a worst-case error will knock out only a single symbol in any row. Since this is the case, we can get away with a single correcting code in each row, and again, no vertical check. The final block is thus 24 symbols across (22 data plus 2 check), and 24 symbols deep. The correction capability is right on the limit of the specification, but the redundancy is now only 11.1%. Remember, there is no requirement that each encoded row contains all $2^m - 1$ symbols so a row of 24 symbols instead of 31 is quite acceptable (although less efficient). Figure 14.3 illustrates this new symbol-wise interleave format.

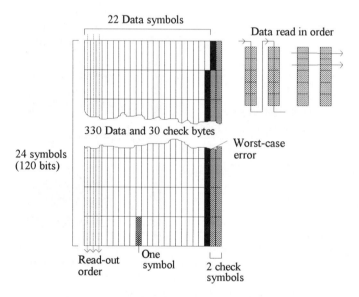

Fig. 14.3 Data arrangement for application two symbol-wise interleave.

In this example it makes little difference whether the data are read into the block as symbols, or as bits (as in Fig. 14.2) to form the rows. Generating the symbols vertically, however, means that much fewer check symbol rows are required giving a reduction in calculations. Having halved the number of correctable errors per row we might expect a halving of the necessary redundancy. In fact the redundancy is now a quarter of that used in the previous example. We also have room over this field for the addition of inner codes if we wish, which would increase the redundancy from 11.1% to around 18%.

In a situation like this one, the coding scheme and the block sizes would be worked out in tandem rather than defining a data block size and trying to fit an error code around it. If we said that the block size was 330 bytes instead of 320, then the redundancies become 27% and 8.3% respectively for the two solutions since we are no longer wasting ten bytes.

The manufacturer has not specified any special requirements which guide the choice of FD or TD either way. The double correcting nature of the first codes suggests perhaps a frequency domain approach but the bit-wise block-interleaving means that there are 60 blocks to code. This workload may be better handled in the time domain. In the second example, time domain coding is the obvious choice since it is trivial for single correcting codes.

It is interesting to note that with bit-wise block-interleaving, in order to satisfy the minimum system requirements, we actually produce something that has well over twice the required capability. We pay for this in terms of redundancy of course, as well as processing. By modifying the shape of the interleaved block, we could squeeze the bit-wise block such that it was 120 bits high, sufficient for a single correcting code per row, but we then have 120 rows of eight symbols to process. Amazingly, even though we can now used a single correcting code in each row, the redundancy is still 27%. The reason for this is because at only eight symbols per row (as opposed to the potential 15) inefficient use of the codes is being made.

Without doubt we would opt for the symbol-wise interleaved solution since this offers very good performance, similar in fact to parity as far as redundancy is concerned, and yet with a very powerful correctional ability.

14.3 SOME REAL APPLICATIONS

To provide a little inspiration, some 'real life' error coding applications are considered here. Possibly one of the most demanding applications of error coding (apart from deep-space communications) has been in RDAT, the consumer digital audio tape system. Tape is fraught with problems such as stretching, dropout and surface damage at the best of times, but the very high packing densities demanded by a fast search (hence short tape) requirement have multiplied them in the case of RDAT.

To give the system an almost random access capability to individual tracks, the tape has to be short, but to maintain useful playing times, the packing density has to be enormous. What might be an almost imperceptible glitch on your average tape, represents literally yards of data in RDAT. Each data track is about 13.5 µm wide, and I estimated the bit size at somewhat under 1.5 µm long. A 0.5 mm² defect translates into about 3.7 cm of track, and a massive 26, 500 bits of channel coded data. The largest consecutive loss for a round defect would be about 570 bits.

To make matters worse there is no space between tracks on the tape, so when the head reads a track it actually picks up substantial parts of both neighbouring tracks. This reduces the signal-to-noise ratio of the system so increasing the probability of random errors. 'Could it ever work?' you might ask, and without error coding the answer is definitely not.

So to what lengths have the designers had to go in order reach an acceptable performance? Cross-interleaving of the data forms a substantial part of the coding and is more complex than I would care to describe here. Tape defects potentially cause very large losses of consecutive data while the reduced signal-to-noise ratio due to crosstalk from adjacent tracks will tend to cause more evenly distributed errors. While interleaving will not help with the latter of these it is an absolutely essential weapon against the former.

With audio there are a number of tricks that can be played to disguise errors in the event that complete correction is not possible, so there is a kind of interleaving within interleaving to assist this **concealment** potential. Essentially an outer code is generated by constructing blocks (from pre-interleaved data) of 32 bytes. These contain 26 data and six Reed–Solomon check symbols. These are interleaved again to form blocks of 32 bytes, but this time with 28 bytes from the outer blocks, and four bytes of inner Reed–Solomon codes.

Errors will be a mix of large bursts due to tape defect plus a distribution of random errors. The inner codes will comfortably iron out the isolated random errors so preventing these from impairing the outer codes which deal with the burst errors. Recalling application two above, you will note that while we can accommodate the 15 bytes of consecutive errors, a single isolated error unfortunately placed, will destroy this capability. So here, you see inner and outer codes complementing each other so as to optimize their capacity.

The compact disc (CD) system uses a very similar approach to RDAT, being affected somewhat by the same problems. The optical geometry of the CD system ensures that small surface defects have little effect on the reading process, so when errors come, they are usually gross. As with RDAT the data is audio, so the interleaving is more complex than might be expected in order to assist concealment in the event of unrecoverable data.

It might be fair to consider a convolutional example of error coding. The CCITT 'V' series of standards, defining communications over the public switched telephone network (PSTN) includes a range of convolutionally encoded (scrambler-based) options (V.22, V.26, V.32, for example). V.32 actually incorporates a convolutional-based error correction option as well. These standards use **modems** (MOdulator/DEModulator) to connect the digital data onto an essentially analogue medium, and the types of data that will be conveyed are quite unknown by the modem manufacturer. For this reason block coding is not an ideal solution because unlike convolutional coding it implies some fore-knowledge of the quantity of data involved.

In rather the same way that convolutional coding fits in well with soft decision decoding, it is also useful as a means of modulating data. Raw data can have properties which are very unsuitable for communication systems. These properties include long runs of ones or zeros, DC components and the like. To get around this, various encoding strategies exist, but convolutional encoders also do the job rather well. Not only this, but the output from the convolutional encoder maps conveniently onto the amplitude and phase modulation scheme (**constellation**) employed by some modems. In a typical scenario, although more data must be transmitted when using the convolutionally encoded message, the coding gain achieved more than offsets this. So in this case the use of convolutional coding, rather than adding complexity to the system, complements it perfectly.

Mobile digital communication is another area which relies heavily upon error correction, often augmented by the ability to retransmit in the case of serious errors. Since much modern digital communication is packet-based having the need to be time domain multiplexed and frequency hop (to allow roaming of the mobile end of the link), block coding lends itself to this application. It has been estimated that a typical mobile link suffers a bit error rate of about one bit in 100. Acceptable errors are in the region of one bit in 1 000 000 so error coding has its work cut out.

The MOBITEX network, developed by Ericsson and Swedish telecom, and in the process of being implemented in about 13 countries, uses (12, 8) Hamming codes extensively. The cellular digital packet data system (CDPD), designed to allow packet data services to overlay onto existing analogue cellular systems, makes use of (63, 47) Reed–Solomon coding. A packet comprises 63 six-bit symbols (378 bits) of which 16 (96 bits) are redundancy. This allows some eight symbol errors per packet, or a worst-case BER of about one bit in 47 (one error bit per symbol) and a best case BER of one in almost eight bits (six error bits per symbol). Digital cellular phone systems such as IS-54, IS-95, GSM and DCS-1800 actually use convolutional coding at rates varying from 1/3 to 1/2. The latency associated with these systems can be quite large owing to the use of speech compression algorithms which can also lead to variable data rates. On this basis, convolutional coding proves quite appropriate, having no need to exacerbate the latency problem, or wait for complete blocks of data.

14.4 SUMMARY OF GROUND COVERED

To close, I should like to round off by summarizing the various ideas that have been covered in the previous chapters. The basic concepts that were introduced early on include the idea of adding redundancy to create a larger difference or space between valid messages than would naturally occur, Hamming distance which is a measure of that space, and orthogonality, a way of generating the space.

On the heels of these basic concepts, we examined their application to error-detection mechanisms such as parity and the cyclic redundancy check. Parity is limited in use, but has niches in low-speed asynchronous links where irregular data cannot be compiled into blocks. The cyclic redundancy check, on the other hand, offers vastly improved performance but at the expense of latency due to the need to compile blocks of data. The cyclic redundancy check provided a convenient introduction to the generation of finite fields using primitive polynomials and we saw early on how the choice of polynomial influenced the effectiveness of the redundancy check.

Error-correction techniques were introduced on similar lines to error detection, extending both parity (in the guise of Hamming codes) and the CRC into single-bit correcting mechanisms. The idea of soft decision decoding was also introduced showing how a knowledge of the probability of a bit being correct could be used to both influence and enhance the outcome of the correction process. With care, soft decision decoding and the addition of expurgation allowed extension of the CRC to double-bit error correction. From this point on, most attention was given to Reed–Solomon (RS) coding using finite fields. Correction was now performed on the basis of symbols rather than bits. Several techniques were examined for generating the necessary RS codes to be appended to the transmitted message. First, we considered time domain RS coding where data is transmitted as is, with the addition of extra codes. The complexities of increasing the capacity of the code to many errors lead to the use of the more complex, but uniform, frequency domain coding of messages where a Fourier-like transform was employed to generate the necessary redundancy, disguising the message in the process.

At this point it was shown how frequency domain processing could be used to calculate the RS symbols in a time domain message and find the errors in a time domain message, yielding considerable reduction in the processing and latency of decoding while maintaining uniform processing.

Packaging of the data was also considered and it was shown how block-interleaving could be used to decorrelate large errors, allowing comparatively modest codes to correct the message. Extending data into several dimensions was also demonstrated, allowing the use of small fields to create large message blocks. This approach is particularly appropriate to hardware solutions where processing scales in numerous ways with field size. At this point convolutional coding was given an airing since this represents a powerful error coding strategy, highly suited to some applications where latency must be minimized.

The final chapters dealt with hardware realization of processing elements for manipulating finite fields, and practical examples of code usage including made-up scenarios and real systems.

It is my principal hope that you have enjoyed this book and so been inspired to investigate and create error coding systems of your own. Also you should now be a little better armed for dealing with more thorough and theoretical texts if you are hungry for a greater depth of understanding. To assist this process, I have included a small bibliography of relevant books which span both theory and real-world applications of coding.

14.5 BIBLIOGRAPHY

Blahut, R.E. (1983) *Theory and Practice of Error Control Codes*, Addison-Wesley, Reading, MA.

Dixon, R.C. (1994) *Spread Spectrum Systems with Commercial Applications*, 3rd edn, John Wiley & Sons Inc., New York.

Gibson, J.D. (1996) *The Mobile Communications Handbook*, CRC Press, Boca Raton, FL.

Lin, S. and Costello, D.J., Jr. (1983) *Error Control Fundamentals and Applications*, Prentice-Hall, Englewood Cliffs, NJ.

Sweeney, P. (1991) *Error Control Coding: An Introduction*, Prentice-Hall, Hemel Hempstead.

Ungerboeck, G. (1987) Trellis coded modulation with redundant signal sets. *IEEE Commun. Mag.* **25** (2), 5–21.

Viterbi, A.J. (1971) Convolutional codes and their performance in communication systems. *IEEE Transactions on Communication Technology*, **19**, 751–772.

Watkinson, J. (1990) *Coding for Digital Recording*, Cambridge University Press, Cambridge.

Webb, W.T. and Hanzo, L. (1994) *Modern Quadrature Amplitude Modulation*, Pentech Press, London.

Wicker, S.B. and Bhargava, V.K. (1994) *Reed–Solomon Codes and their Applications*, IEEE Press, NJ.

Ziemer, R.E. and Peterson, R.L. (1992) *Introduction to Digital Communication*, Macmillan, New York.

Part Seven

Exercises

15

Questions

1.   Find the weights of the following seven-bit words:

      (a) 1011100, (b) 1100101, (c) 0010110, (d) 0101100, (e) 1101111

2.   Generate an even parity bit for the seven-bit words of question 1.
      The seven-bit words and one-bit parity are to be sandwiched
      between a start and stop bit to form a character suitable for
      transmission. What is the message redundancy and what percentage
      of this is due to the parity bit?

3.   If the data of question 2 are to be compiled into blocks of 16 seven-
      bit words and a vertical checksum added, what is the overall
      message efficiency?

4.   The following message is encoded with even horizontal and vertical
      parity and each data word is eight bits. Correct the message.

      01101000100010011011010010010111001100010100

5.   Using a Venn diagram, demonstrate how a (7, 4) Hamming coded
      message can be constructed.

6.   Using the parity generator equations, $P_1 = a + c + d$, $P_2 = a + b + c$
      and $P_3 = b + c + d$, demonstrate the orthogonal relationship between
      the data and parity with a simple sketch.

7.   The following (16, 11) Hamming encoded message arrives cor-
      rupted. The bits are in the order *dddddddpdddpdpppP* where *d* are

data, p are parity and P is an overall even parity bit:

0110011000101010

Demonstrate that an odd number of bit errors have occurred in the message. Assuming that only a single-bit error is present, correct the message. What is d_{min} for this code?

15.2 CYCLIC REDUNDANCY CHECKING

1. Using the generator polynomial $x^3 + x + 1$, how many bits will be required for the CRC that is produced?

2. Using the polynomial of question 1, in a typical message, what percentage of all errors will be detected?

3. Using $x^3 + x + 1$ for the generator polynomial, find the CRC for the following data: 011010110. What should the maximum size of the complete message be if single-bit error correction is to be possible?

4. The data of question 3 arrive at the receiver with a double-bit error such that it appears 110010110*CRC* where *CRC* is your appended CRC. Show that the error is detected. Would it have been detected if a further bit error occurred such that it was 110110110*CRC*?

5. Demonstrate that the polynomial $x^4 + x + 1 = 0$ is primitive.

6. Using $x^4 + x + 1 = 0$ for the GP, construct the (12, 8) message from the data 10011011. What is d_{min} for this message?

7. Using the polynomial of question 6, a (12, 8) message arrives at the receiver as 101010010101. Demonstrate that more than a one-bit error has occurred. What property of this code allows you to make this judgement, and is it a general result? Demonstrate your answer with an example.

8. You know that the data of question 7 were supposed to be an ASCII text code, and that the most significant bit was probably 0. Assuming

that only two bit errors have occurred, correct the message.

9. Using the polynomial 19 as a starting point, create a GP that is suitable for generating (15, 10) codes. What special property does this code have?

10. Correct the following (15, 10) code, which has been based upon the polynomial 19: 101101111011011.

11. Using a soft decision decoder, another (15, 10) message arrives as 325226145727210 where $7 =$ good 1 and $0 =$ good 0. Assuming there are no more than two bit errors, correct the message.

12. Starting with $x + 1$, generate the sequence of values not represented by the generator polynomial used in questions 9, 10 and 11. Comment on the non-zero value that is not represented in either this sequence or the sequence generated when starting with x.

13. Design a hardware CRC generator checker for an expurgated cyclic code based upon the polynomial $x^4 + x^3 + 1$.

14. Using long division (modulo-2), demonstrate that the CCITT generator polynomial $x^{16} + x^{12} + x^5 + 1$ generates expurgated codes by revealing its two factors.

15.3 MANIPULATING FINITE FIELD ELEMENTS

1. Show that the polynomial $x^2 + x + 1$ is primitive by generating the non-zero elements of the field GF(2^2) from it.

2. Using the starting conditions

$$A + P + Q = 0$$

and

$$A\alpha^0 + P\alpha + Q\alpha^2 = 0$$

where A is a data symbol and P and Q are check symbols, show how P and Q are derived from A for the field GF(2^2).

3.   A message encoded on the basis of question 2 arrives at the receiver as 100101 (*APQ*). Check the message for errors and, if necessary, correct it.

4.   What is the message efficiency in questions 2 and 3? What would this become if the message was extended into two dimensions?

5.   By creating the four possible valid messages of question 2, find d_{min} for this code.

6.   Using the field GF(2^2), calculate the inverse Fourier transform of the data 00, 00, 10 and thus create an error coded message.

7.   A message similarly encoded to that of question 6, arrives at the receiver as 111110. Using full frequency domain decoding, correct the message and find the original data.

8.   A double symbol correcting time domain message has been set up over GF(2^3), and upon examining the non-zero syndromes, you arrive at the quadratic $\alpha^{2x} + \alpha^{4+x} + \alpha^2 = 0$, where x points to the two error positions. Using analytical means, find the two error positions if the primitive polynomial is 11.

9.   By constructing a table, verify that your analytical solutions to question 8 are correct.

10.   A 49-symbol, two-dimensional frequency encoded message based on the polynomial 11, arrives at the receiver corrupted. After Fourier transforming, the message appears as

$$
\begin{array}{cccccccc}
3 & 5 & 3 & 2 & 6 & . & . \\
1 & 6 & 1 & 4 & . & . & \\
3 & 1 & 5 & 6 & . & . & \\
2 & 3 & 4 & . & . & & \\
2 & 5 & . & . & e & t & c \\
. & . & . & & & & \\
. & & & & & &
\end{array}
$$

Assuming that a single error has occurred, identify the error.

11. If the message in question 10 had been set up as a double correcting code, and two symbol errors had occurred, find the rows and columns of the time domain message affected by the errors.

12. Given the starting conditions

$$A + B + C + D + E + P + Q = 0$$

and

$$A\alpha^0 + B\alpha^1 + C\alpha^2 + D\alpha^3 + E\alpha^4 + P\alpha^5 + Q\alpha^6 = 0$$

are used to force two roots into an error encoded message, generate expressions to find the check symbols P and Q based on the data symbols A to E. Use the polynomial 11 over GF(2^3).

13. The message 1527531 (A–Q) arrives at a receiver having been encoded as in question 12. Assuming only single symbol error, correct the message.

14. A second message, 4152323, arrives, encoded in the same way as question 13, but this time two errors have occurred. One error is in symbol position 0, while the other error has the bit pattern 3. Correct the message.

15. Using the scheme of question 12, the following time domain encoded message block has both single symbol inner and outer codes appended to it. Identify the rows and columns in error and comment on your findings. See if you can correct the entire message.

1	5	4	7	5	3	1
7	1	5	2	0	2	3
6	3	7	2	1	1	3
7	5	7	6	2	5	0
7	6	4	5	2	7	3
6	2	2	4	3	1	0
6	6	3	4	6	3	2

16. In question 12, you generated two equations for calculating P and Q based on A to E. Using a process a little like erasure, show how P and Q may be found from two syndromes, S_0 and S_1, which result when P and Q are initially set to 0.

17. Use the techniques of questions 12 and 16 to find P and Q for the data $(A-E)$, 03365, thereby demonstrating that you get the same answer.

18. Generate a matrix solution for question 16, and use it to evaluate P and Q for the data of question 17.

19. For the data of question 17, use the Fourier transform method to find the check symbols P and Q.

20. You are required to generate a block code for 1024 bits of data. What is the minimum field size required to encode this message into a single one-dimensional block with single symbol correcting capability? Calculate the minimum field sizes for two-, three- and four-dimensional data structures.

21. What is the message redundancy in each case in question 20?

15.4 CONVOLUTIONAL CODING

1. Draw out the trellis diagram for the following circuit:

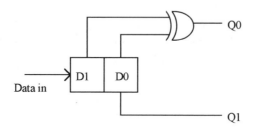

What is the coding rate for this encoder?

2. Using the encoder of question 1, generate the encoded data stream for the message 01001011.

3. A message encoded using the circuit of question 1 arrives as (Q1/Q0), 00 01 00 10 11 10 11 00 01. Find the most likely source data.

4. The received data stream of question 3 is available as probabilities where 0 is a high probability of 0, and 7 is a high probability of 1. The message before thresholding was 10 26 33 62 56 43 77 21 14. Correct the message now.

5. From the trellis, derive the encoder circuit

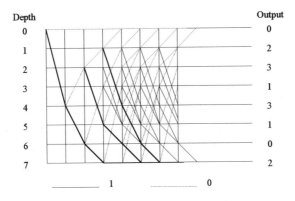

16

Solutions

16.1 PARITY

1. 4, 4, 3, 3, 6.

2. 0, 0, 1, 1, 0. Since three out of every ten bits are not information, the redundancy is 30%. Ten per cent of the message redundancy is due to parity (one bit in ten).

3. Each block comprises 135 bits $(16 \times 8 + 7)$ of which 16×7 (112) are information. The efficiency is, therefore, $100 \times 112/135 = 83\%$.

4. Compiling the data into a block gives

0	1	1	0	1	0	0	0	1	✓
0	0	0	1	0	0	1	1	0	✗
1	1	0	1	0	0	1	0	0	✓
1	0	1	1	1	0	0	1	1	✓
0	0	0	**1**	0	1	0	0	0	✓
✓	✓	✓	✓	✓	✗	✓	✓		

    The bold **0** intersects with both horizontal and vertical parity errors. Assuming that there is only a single error, this bit should be inverted.

5.

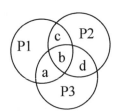

The data, *a, b, c* and *d* are arranged in the intersecting regions of the circles while the parity bits fill the remainder of the unused spaces. Each parity bit is set or cleared to ensure a known parity within its own circle.

6. The data are arranged on orthogonal axes, with the surface bounded by the axes being the respective parity. *c* is used in the generation of all parity bits and therefore sits at the origin since this point is included in all surfaces.

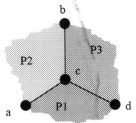

7. The weight of the message is seven, so the overall parity of the message is incorrect. This error must be the result of an odd number of bits. $p8$ is calculated on the basis of the left-hand eight bits so

$$01100110xxxxxxxx, \text{ no error} = 0$$

$p4$ comes from
$$0110xxxx0010xxxx, \text{ error} = 1$$

$p2$ comes from
$$01xx01xx00xx10xx, \text{ error} = 1$$

$p1$ comes from
$$0x1x0x1x0x1x1x1x, \text{ error} = 1$$

The syndrome is 0111, pointing to bit seven. Remember that the right-most parity bit is extra to the Hamming code so the corrected message is 0110011010101010. d_{min} for the basic code will be three, but the addition of an overall parity bit increases this to four.

16.2 CYCLIC REDUNDANCY CHECKING

1. The given polynomial is four bits (1011) so the CRC will require three bits.

2. Seven in eight, or 88%, of all errors will be detected.

3. The CRC is 111. Since there are three CRC bits, the message should not exceed seven bits if a single bit is to be correctable.

```
1  0  1  1 | 0  1  1  0  1  0  1  1  0      0  0  0
             1  0  1  1
             1  1  0  0
             1  0  1  1
                1  1  1  1
                1  0  1  1
                   1  0  0  1
                   1  0  1  1
                      1  0  0        0
                      1  0  1        1
                         1           1  0  0
                         1           0  1  1
                         0          [1  1  1]
```

4. The non-zero remainder 010 shows us that the message has been corrupted. A further bit error in the position suggested would lead to an error of the form 1011. Since this is equal to the GP, it would not be detected.

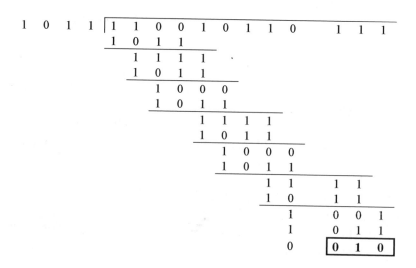

```
1  0  1  1 | 1  1  0  0  1  0  1  1  0      1  1  1
             1  0  1  1
             1  1  1  1
             1  0  1  1
                1  0  0  0
                1  0  1  1
                   1  1  1  1
                   1  0  1  1
                      1  0  0  0
                      1  0  1  1
                         1  1     1  1
                         1  0     1  1
                            1     0  0  1
                            1     0  1  1
                            0    [0  1  0]
```

5. Using x as the primitive root, then consistently multiplying by x yields the following sequence:

$$x, x^2, x^3, x + 1, x^2 + x, x^3 + x^2, x^3 + x + 1, x^2 + 1, x^3 + x,$$

$$x^2 + x + 1, x^3 + x^2 + x, x^3 + x^2 + x + 1, x^3 + x^2 + 1, x^3 + 1, 1, x, \text{ etc.}$$

which numerically is

$$2, 4, 8, 3, 6, 12, 11, 5, 10, 7, 14, 15, 13, 9, 1, 2$$

Since this is the sequence of all non-zero elements of four bits, we know that $x^4 + x + 1$ is primitive.

6. The (12, 8) message is, therefore, 100110110101. Because this is a single-bit correcting code, d_{min} must be three.

```
1 0 0 1 1 | 1 0 0 1 1 0 1 1   0 0 0 0
            1 0 0 1 1
                          1 1   0 0 0
                          1 0   0 1 1
                             1   0 1 1 0
                             1   0 0 1 1
                                 0 1 0 1
```

7.

```
1 0 0 1 1 | 1 0 1 0 1 0 0 1   0 1 0 1
            1 0 0 1 1
              1 1 0 0 0
              1 0 0 1 1
                1 0 1 1 1
                1 0 0 1 1
                    1 0 0   0 1
                    1 0 0   1 1
                           1 0 0 1
```

The remainder 9 corresponds to α^{14} from the list generated in question 5. Since valid error bit positions within the message range from 0 to 11, an error in bit 14 is impossible. Clearly more than a single-bit error must have occurred. Because the code is shortened, we can sometimes make this deduction, but this is not a general result. Take for example errors in bit positions 0 and 1. They would appear as a single error in bit four. Because bit errors in the form of the GP (10011) go unnoticed, it follows that bit errors in two adjacent bit positions n and $n + 1$ will look like an error at position $n + 4$.

8. The error syndrome was nine. An error in the most significant bit alone would result in a syndrome of α^{11}, so the remaining error will be

$$9 + \alpha^{11} = 9 + 14 = 7 \text{ or } \alpha^{10}$$

The corrected message is **011010010101**.

9. A (15, 10) code has five redundancy bits added. Since the GP 19 will create only four redundancy bits, the code must be expurgated. The modified GP is, therefore, $(x + 1)(x^4 + x + 1) = x^5 + x^4 + x^2 + 1$. Using this generator polynomial, the (15, 10) code has a d_{min} of four rather than three. This means that single- and double-bit errors can be distinguished.

10.

```
1 1 0 1 0 1 | 1 0 1 1 0 1 1 1 1 0 1 1 0 1 0
              1 1 0 1 0 1
                1 1 0 0 0 1
                1 1 0 1 0 1
                    1 0 0 1 1 0
                    1 1 0 1 0 1
                    0 1 0 0 1 1 1
                      1 1 0 1 0 1
                        1 0 0 1 0 1
                        1 1 0 1 0 1
                          1 0 0 0 0 0
                          1 1 0 1 0 1
                            1 0 1 0 0 1
                            1 1 0 1 0 1
                              1 1 1 0 0 0
                              1 1 0 1 0 1
                                0 1 1 0 1
```

The received message generated the syndrome 01101, so an error has obviously occurred. To correct the error, we must complete the finite field using the generator polynomial $x^5 + x^4 + x^2 + 1 = 0$.

Starting with x we get:

$$x, x^2, x^3, x^4, x^4 + x^2 + 1, x^4 + x^3 + x^2 + x^1 + 1, x^3 + x^1 + 1,$$
$$x^4 + x^2 + x^1, x^4 + x^3 + 1, x^2 + x^1 + 1, \ x^3 + x^2 + x^1,$$
$$x^4 + x^3 + x^2, x^3 + x^2 + 1, x^4 + x^3 + x^1, 1, x, x^2, \text{etc.}$$

Numerically this is

2, 4, 8, 16, 21, 31, 11, 22, 25, 7, 14, 28, 13, 26, 1, 2, 4, etc.

The remainder was 13 or 01101, which corresponds to α^{13} from the above list. The corrected message is, therefore, 111101111011010.

11. Repeating, but with the (thresholded data) 001001011101000, we get the remainder

```
1 1 0 1 0 1 | 0 0 1 0 0 1 0 1 1 1 0 1 0 0 0
              1 1 0 1 0 1
              _____
              1 0 0 0 0 1
              1 1 0 1 0 1
                1 0 1 0 0 1
                1 1 0 1 0 1
                _____
                  1 1 1 0 0 0
                  1 1 0 1 0 1
                  _____
                    1 1 0 1 1 0
                    1 1 0 1 0 1
                    _____
                    [0 1 1 0 0]
```

The remainder 12 does not appear in the list of question 10. It must be the result of two or more bit errors. Assuming a two-bit error, the bit pairs which can give rise to a remainder of 12 are

$$\alpha^0/\alpha^{13}, \ \alpha^1/\alpha^{11}, \ \alpha^2/\alpha^3, \ \alpha^4/\alpha^{12}, \ \alpha^5/\alpha^9, \ \alpha^7/\alpha^{10}, \ \alpha^8/\alpha^{14}$$

We must now consider the combined probabilities of these bit pairs being correct.

Index	E	D	C	B	A	9	8	7	6	5	4	3	2	1	0
Code	3	2	5	2	2	6	1	4	5	7	2	7	2	1	0
Prob.	1	2	2	2	2	3	3	1	2	4	2	4	2	3	4

Pair	0/13	1/11	2/3	4/12	5/9	7/10	8/14
Prob.	16	6	8	4	12	2	3

The bit pair 7/10 exhibits the lowest combined probability of being correct and, therefore, this is the pair we choose. The corrected message is 0010**1**10011**0**1000.

12. $x + 1$, $x^2 + x$, $x^3 + x^2$, $x^4 + x^3$, $x^2 + 1$, $x^3 + x$, $x^4 + x^2$, $x^4 + x^3 + x^2 + 1$,
$x^3 + x^2 + x + 1$, $x^4 + x^3 + x^2 + x$, $x^3 + 1$, $x^4 + x$,
$x^4 + 1$, $x^4 + x^2 + x + 1$, $x^4 + x^3 + x + 1$, $x + 1$

or

3, 6, 12, 24, 5, 10, 20, 29, 15, 30, 9, 18, 17, 23, 27, 3, ...

The non-zero value not represented in either this sequence or the sequence generated in question 10, is 19, which corresponds to the pattern of the original generator polynomial (before multiplication by $x + 1$).

13. Multiplying $x^4 + x^3 + 1$ by $x + 1$ gives $x^5 + x^3 + x + 1$ so the CRC generator/checker circuit will have the following form:

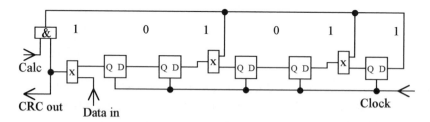

14. From the calculation below we can see that $(x + 1)$ divides into the GP since there is no remainder. The GP will, therefore, generate expurgated codes. The other factor is found from the top line of the division.

```
              1 1 1 1 0 0 0 0 0 0 0 1 1 1 1 1
    1  1 | 1  0  0  0 1 0 0 0 0 0 0 1 0 0 0 0 1
         1  1
         1  0
         1  1
            1  0
            1  1
               1  1
               1  1
               0  0 0 0 0 0 0 0 1 0
                               1  1
                                  1  0
                                  1  1
                                     1  0
                                     1  1
                                        1  0
                                        1  1
                                           1  1
                                           1  1
```

$$x^{15} + x^{14} + x^{13} + x^{12} + x^4 + x^3 + x^2 + x^1 + 1$$

16.3 MANIPULATING FINITE FIELD ELEMENTS

1. Starting with x, we get the sequence

Power	Calculation	Decimal
α^1	x	2
α^2	$x.x = x^2 = x + 1$	3
$\alpha^3 = \alpha^0$	$(x+1).x = x^2 + x = 1$	1
α	$1.x = x$	2
	etc.	

Since all the non-zero elements are represented by this sequence, the polynomial must be primitive.

2. $P = A + Q$ so substituting into $A\alpha^0 + P\alpha^1 + Q\alpha^2 = 0$ gives

$$A(\alpha^0 + \alpha) + Q(\alpha + \alpha^2) = 0$$
$$A(1 + \alpha) + Q\alpha(1 + \alpha) = 0$$

$$Q = A\alpha^{-1} = A\alpha^2$$

and so

$$P = A + A\alpha^2 = A\alpha$$

3.
$$S_0 = A + P + Q \qquad = 10 + 01 + 01 = 10\ (\alpha)$$
$$S_1 = A + P\alpha + Q\alpha^2 = 10 + 10 + 11 = 11\ (\alpha^2)$$

$$k = S_1/S_0 = \alpha^1$$

so the error symbol is position 1 or P.

$$P = 01 + S_0 = 01 + 10 = 11$$

so the corrected message is
$$10\ 11\ 01$$

4.   The message efficiency or data bits/message bits is 33%. If the message was extended into two dimensions then we would have nine two-bit symbols of which three were redundant. This leads to an efficiency of 6/9 or 67%.

5.   Since there is one two-bit data symbol in the message, there will be four valid messages as follows:

$$
\begin{array}{llll}
A_0 = 00\ (0)   & P_0 = 00 & Q_0 = 00 & = 00\ 00\ 00 \\
A_1 = 01\ (\alpha^0) & P_1 = 10 & Q_1 = 11 & = 01\ 10\ 11 \\
A_2 = 10\ (\alpha^1) & P_2 = 11 & Q_2 = 01 & = 10\ 11\ 01 \\
A_3 = 11\ (\alpha^2) & P_3 = 01 & Q_3 = 10 & = 11\ 01\ 10
\end{array}
$$

Comparing the four valid messages, we find that four bits must be changed to get from any one code to any other. d_{min} is, therefore, 4.

6.
$$d_0 = F_0 + F_1 + F_2$$

$$d_1 = F_0\alpha^0 + F_1\alpha^{-1} + F_2\alpha^{-2} = F_0\alpha^0 + F_1\alpha^2 + F_2\alpha^1$$

$$d_2 = F_0\alpha^0 + F_1\alpha^{-2} + F_2\alpha^{-4} = F_0\alpha^0 + F_1\alpha^1 + F_2\alpha^2$$

$$d_0 = F_2 = 10,\ d_1 = F_2\alpha^1 = 11,\ d_2 = F_2\alpha^2 = 01$$

so the message is

$$10\ 11\ 01$$

7. Calculating the FFT on the data 11 11 10 gives

$$F_0 = \alpha^2 + \alpha^2 + \alpha = \alpha$$
$$F_1 = \alpha^2 + \alpha^0 + \alpha^0 = \alpha^2$$

Since these are not zero, and they should be, we know that an error has occurred so there is no point in calculating F_2. We must complete the error spectrum so

$$L_1 = F_1/F_0 = \alpha$$
$$F_2 = L_1F_1 = \alpha^0$$

We now need the IFT of the completed error spectrum in order to find the time domain error pattern. This gives 0, α, 0 or 00 10 00, so the corrected message is 11 01 10. To find the original data we must perform an FFT on this. However, F_0 and F_1 should be 0 so (apart from to check) we need only calculate F_2 which contains the data.

$$\text{Data} = 00\ 00\ 11$$

8. Using the primitive polynomial 11, from the notes we know that the solution to a quadratic is given by

$$a_2(a_0 + p_0) + a_1(a_1 + p_1) + (a_2 + a_0)(a_2 + p_2) = q_2$$
$$a_1(a_0 + p_0) + (a_2 + a_0)(a_1 + p_1) + (a_2 + a_1)(a_2 + p_2) = q_1$$
$$a_0(a_0 + p_0) + a_2(a_1 + p_1) + a_1(a_2 + p_2) = q_0$$

In this case, $p = 110\ (\alpha^4)$ and $q = 100\ (\alpha^2)$ so

$$a_2(a_0 + 0) + a_1(a_1 + 1) + (a_2 + a_0)(a_2 + 1) = 1$$
$$a_1(a_0 + 0) + (a_2 + a_0)(a_1 + 1) + (a_2 + a_1)(a_2 + 1) = 0$$
$$a_0(a_0 + 0) + a_2(a_1 + 1) + a_1(a_2 + 1) = 0$$

From the first of these,

$$a_0 a_2 + a_1 a_1 + a_1 + a_2 a_2 + a_2 + a_0 a_2 + a_0 = 1$$
$$a_0 = 1$$

From the second,

$$a_0 a_1 + a_1 a_2 + a_0 a_1 + a_2 + a_0 + a_2 a_2 + a_2 + a_1 a_2 + a_1 = 0$$
$$a_0 + a_1 + a_2 = 0$$
$$1 + a_1 + a_2 = 0$$
$$a_1 = \bar{a}_2$$

From the last

$$a_0(a_0 + 0) + a_2(a_1 + 1) + a_1(a_2 + 1) = 0$$
$$a_0 a_0 + a_1 a_2 + a_2 + a_1 a_2 + a_1 = 0$$

Same result as previous manipulation. From these, we can produce two results, 011 and 101 since a_0 is always 1 and $a_1 = \bar{a}_2$. This means that the error symbols are in positions 3 and 6, corresponding to α^3 and α^6 derived above.

9.

Power	Calculation	Decimal	=0?
0	$\alpha^0 + \alpha^4 + \alpha^2$	3	✗
1	$\alpha^2 + \alpha^5 + \alpha^2$	7	✗
2	$\alpha^4 + \alpha^6 + \alpha^2$	7	✗
3	$\alpha^6 + \alpha^7 + \alpha^2$	0	✓
4	$\alpha^1 + \alpha^1 + \alpha^2$	4	✗
5	$\alpha^3 + \alpha^2 + \alpha^2$	3	✗
6	$\alpha^5 + \alpha^3 + \alpha^2$	0	✓

10. For a single error, we know that the three top left corner elements should have been zero. Since they aren't, we can use them to find the error. $L_x = \alpha^6/\alpha^3 = \alpha^3$ and $L_y = \alpha^3/\alpha^3 = \alpha^0$. For a single correcting code, these directly betray the error location at position (3, 0).

The corner spectra is $\xi\alpha^{ij}$ where i and j are the error ordinates and ξ is the error pattern, so

$$\xi = \alpha^3/\alpha^{-ij}$$
$$\xi = \alpha^7 \text{ or } 001$$

11. This time we need to find L_1 and L_2. The solution to L_1 is given by

$$L_{x1} = (E_0E_3 + E_1E_2)/(E_1E_1 + E_0E_2)$$

so

$$L_{x1} = (\alpha^{3+1} + \alpha^{6+3})/(\alpha^{6+6} + \alpha^{3+3}) = \alpha/\alpha = \alpha^0$$

$$L_{x2} = (E_1E_3 + E_2E_2)/(E_1E_1 + E_0E_2)$$

so

$$L_{x2} = (\alpha^{6+1} + \alpha^{3+3})/(\alpha^{6+6} + \alpha^{3+3}) = \alpha^2/\alpha = \alpha^1$$

From this we can construct the quadratic

$$\alpha^{2x} + \alpha^x + \alpha = 0$$

Solving gives $x = 2$ and $x = 6$, the columns affected.

$$L_{y1} = (E_0E_3 + E_1E_2)/(E_1E_1 + E_0E_2)$$

so

$$L_{y1} = (\alpha^{3+1} + \alpha^0\alpha^3)/(\alpha^{0+0} + \alpha^3\alpha^3) = \alpha^6/\alpha^2 = \alpha^4$$

$$L_{y2} = (E_1E_3 + E_2E_2)/(E_1E_1 + E_0E_2)$$

so

$$L_{y2} = (\alpha^{0+1} + \alpha^{3+3})/(\alpha^{0+0} + \alpha^{3+3}) = \alpha^5/\alpha^2 = \alpha^3$$

From this we get the quadratic

$$\alpha^{2y} + \alpha^{y+4} + \alpha^3 = 0$$

and solving gives $y = 1$ and $y = 2$, the rows affected.

12. $Q = A + B + C + D + E + P$, so substituting into the second equation gives

$$A(\alpha^0 + \alpha^6) + B(\alpha + \alpha^6) + C(\alpha^2 + \alpha^6) + D(\alpha^3 + \alpha^6)$$
$$+ E(\alpha^4 + \alpha^6) + P(\alpha^5 + \alpha^6) = 0$$

so

$$A\alpha^2 + B\alpha^5 + C\alpha^0 + D\alpha^4 + E\alpha^3 + P\alpha^1 = 0$$

so

$$\boldsymbol{P = A\alpha^1 + B\alpha^4 + C\alpha^6 + D\alpha^3 + E\alpha^2}$$

and

$$Q = A(1 + \alpha^1) + B(1 + \alpha^4) + C(1 + \alpha^6) + D(1 + \alpha^3) + E(1 + \alpha^2)$$

$$\boldsymbol{Q = A\alpha^3 + B\alpha^5 + C\alpha^2 + D\alpha + E\alpha^6}$$

13. Calculating the syndromes from the data

	Calculate S_0			Calculate S_1		
Power	Binary	Dec.	Power	Binary	Dec.
α^0	001	1	$\alpha^0\alpha^0$	001	1
α^6	101	5	$\alpha^6\alpha^1$	001	1
α^1	010	2	$\alpha^1\alpha^2$	011	3
α^5	111	7	$\alpha^5\alpha^3$	010	2
α^6	101	5	$\alpha^6\alpha^4$	011	3
α^3	011	3	$\alpha^3\alpha^5$	010	2
α^0	001	1	$\alpha^0\alpha^6$	101	5
S_0	110	6	S_1	101	5

From this, S_1 is α^6 and S_0 is α^4 so the error location is

$$k = \alpha^6/\alpha^4 = \alpha^2 \text{ or element } C \text{ and the error is } 110 \ (S_0) \text{ so}$$

$$C = 010 + 110 = 100$$

so the corrected message is

$$1547531 \ (A - Q)$$

14. Since two errors have occurred, $S_0 = \xi_i + \xi_j$ and $S_1 = \xi_i \alpha^i + \xi_j \alpha^j$. Now we know that $i = 0$ (the position of one of the errors) and $\xi_j = 3$ (the pattern of the other error).

Calculating the syndromes, $S_0 = \alpha^2$ and $S_1 = \alpha^6$ so

$$\alpha^2 = \xi_0 + \alpha^3 \text{ and } \alpha^6 = \xi_0\alpha^0 + \alpha^3\alpha^j$$

$$\xi_0 = \alpha^2 + \alpha^3 = \alpha^5 \text{ and } \alpha^6 = \alpha^5\alpha^0 + \alpha^3\alpha^j \text{ so } \alpha^j = \alpha/\alpha^3 = \alpha^5$$

The error pattern of A is 111 so $A = 011$, and the index of the known error pattern, 011, is 5 (or P) so $P = 001$, the corrected message is

$$3152313$$

15. Calculating syndromes horizontally and vertically, we find that columns 2, 3 and 4 exhibit errors (starting at 0 on the left) and columns 2, 3 and 4 exhibit errors (top row 0). What may be surprising is that rows 5 and 6 give zero syndrome even though the check symbols 1, 0, 3, 2 were calculated in columns.

We can forget attempting a single-symbol correction since if we attempt a double correction it will also repair a single error. For each row or column, there are three possible combinations of two-bit errors, 2/3, 2/4 and 3/4, giving rise to 18 solutions. One possible approach to this situation is using erasure to perform the three possible corrections on row 2. If two of these give similar results, then either there is only a single error on this row, or there is a three-symbol error. Having 'corrected' this row, then erasure is used on the three solutions to correct the columns assuming errors in rows 3 and 4. This process is repeated, correcting row 3 horizontally, and the three columns based on errors in rows 2 and 4. Finally, row 4 is corrected horizontally and the columns completed on the basis of errors in rows 2 and 3. This gives nine possible solutions.

The operation may be repeated, correcting the columns before rows. There are situations where the capacity in all rows may be exceeded, but columns may still be correctable, allowing subsequent correction of the rows. This yields a further nine solutions. Having attempted the correction, the syndromes can be checked in both directions for all solutions, discarding any that may be non-zero. In this example, several of the solutions are the same. We must be careful not to assume that this makes them more likely to be the correct solution. There are six different solutions for this question as follows (showing only the rows and columns in error).

6	0	1
6	6	7
0	3	6

0	5	2
3	4	0
3	4	2

5	7	5
1	7	1
4	5	4

1	1	7
7	2	2
6	6	5

3	2	6
4	5	6
7	2	0

7	4	4
2	0	5
5	1	1

We might use soft decision decoding to help us make the final choice, or perhaps a knowledge of the context of the data may provide a clue. The original data was 7, 4, 4, 2, 0, 5, 5, 1, 1.

16. If the check symbols are initially zero, then

$$S_0 = P + Q \quad \text{and} \quad S_1 = \alpha^5 P + \alpha^6 Q$$

Solving gives $P = \alpha^6 S_1 + \alpha^5 S_0$ and $Q = \alpha^6 S_1 + \alpha^4 S_0$.

17. Using the equations from solution 12 we have

$$P = A\alpha^1 + B\alpha^4 + C\alpha^6 + D\alpha^3 + E\alpha^2$$

and

$$Q = A\alpha^3 + B\alpha^5 + C\alpha^2 + D\alpha + E\alpha^6$$

so using the data 0, α^3, α^3, α^4, α^6, we get

$$P = 0\alpha^1 + \alpha^3\alpha^4 + \alpha^3\alpha^6 + \alpha^4\alpha^3 + \alpha^6\alpha^2$$
$$= 001 + 100 + 001 + 010 = \alpha^4$$

$$Q = 0\alpha^3 + \alpha^3\alpha^5 + \alpha^3\alpha^2 + \alpha^4\alpha + \alpha^6\alpha^6$$
$$= 010 + 111 + 111 + 111 = \alpha^6$$

Using the solution to question 16,

$$S_0 = 000 + 011 + 011 + 110 + 101 = 011$$

and

$$S_1 = 000 + 110 + 111 + 001 + 011 = 011$$

so

$$P = \alpha^6\alpha^3 + \alpha^5\alpha^3 = 100 + 010 = \alpha^4$$

and

$$Q = \alpha^6\alpha^3 + \alpha^4\alpha^3 = 100 + 001 = \alpha^6$$

18. Arranging $S_0 = P + Q$ and $S_1 = \alpha^5 P + \alpha^6 Q$ into a matrix format gives

$$\begin{bmatrix} 1 & 1 \\ \alpha^5 & \alpha^6 \end{bmatrix} \begin{bmatrix} P \\ Q \end{bmatrix} = \begin{bmatrix} S_0 \\ S_1 \end{bmatrix}$$

Inverting the matrix gives

$$\frac{1}{\Delta}\begin{bmatrix} \alpha^6 & 1 \\ \alpha^5 & 1 \end{bmatrix}\begin{bmatrix} S_0 \\ S_1 \end{bmatrix} = \begin{bmatrix} P \\ Q \end{bmatrix}$$

where $\Delta = 1.\alpha^5 + 1.\alpha^6 = \alpha$, so

$$\begin{bmatrix} \alpha^5 & \alpha^6 \\ \alpha^4 & \alpha^6 \end{bmatrix}\begin{bmatrix} S_0 \\ S_1 \end{bmatrix} = \begin{bmatrix} P \\ Q \end{bmatrix}$$

It is obvious from here that we will obtain the same results as before.

19. We know that if P and Q were set correctly, F_0 and F_1 for the message would be 0. These frequency components correspond to the two syndromes, so we don't need to recalculate them for the initial case where P and Q are set to zero, i.e.

$$F_0 = \alpha^3 \text{ and } F_1 = \alpha^3$$

Our known error positions are 5 and 6, the locations of the two check symbols, P and Q, so $L_1 = \alpha^5 + \alpha^6 = \alpha^1$ and $L_2 = \alpha^{5+6} = \alpha^4$. Using these results we can set up an REC to calculate the check symbol spectrum of P and Q.

$$F_2 = L_2 F_0 + L_1 F_1 = \alpha^4 \alpha^3 + \alpha^1 \alpha^3 = \alpha^5$$
$$F_3 = L_2 F_1 + L_1 F_2 = \alpha^4 \alpha^3 + \alpha^1 \alpha^5 = \alpha^2$$
$$F_4 = L_2 F_2 + L_1 F_3 = \alpha^4 \alpha^5 + \alpha^1 \alpha^2 = \alpha^5$$
$$F_5 = L_2 F_3 + L_1 F_4 = \alpha^4 \alpha^2 + \alpha^1 \alpha^5 = 0$$
$$F_6 = L_2 F_4 + L_1 F_5 = \alpha^4 \alpha^5 + \alpha^1 0 = \alpha^2$$

To find P and Q we must IFT this spectrum. The PQ spectrum $= \alpha^3$, α^3, α^5, α^2, α^5, 0, α^2. Since only P and Q should be non-zero, we need not calculate the whole spectrum, but for fun we'll find d_0.

$$d_0 = 011 + 011 + 111 + 100 + 111 + 000 + 100 = 0$$

$$d_5 (P) = \alpha^3 \alpha^0 + \alpha^3 \alpha^2 + \alpha^5 \alpha^4 + \alpha^2 \alpha^6 + \alpha^5 \alpha^1 + 0\alpha^3 + \alpha^2 \alpha^5$$
$$= 011 + 111 + 100 + 010 + 101 + 000 + 001 = 110 = \alpha^4$$

$$d_6 (Q) = \alpha^3\alpha^0 + \alpha^3\alpha^1 + \alpha^5\alpha^2 + \alpha^2\alpha^3 + \alpha^5\alpha^4 + 0\alpha^5 + \alpha^2\alpha^6$$
$$= 011 + 110 + 001 + 111 + 100 + 000 + 010 = 101 = \alpha^6$$

20.   Consider GF(2⁷). This field supports blocks up to 127×7 bits or 889, so it will not be large enough. The smallest field suitable will, therefore, be GF(2⁸). This will support messages up to 2040 bits.

If we go to two dimensions, then GF(2⁵) will support messages up to 4805 bits. GF(2⁴) is just too small at 900 bits. For three-dimensional structures, we can use GF(2⁴). GF(2³) is just a few bits too small by the time the check symbols are added (1029 bits maximum). Over four dimensions, GF(2³) is fine, with a message size of 7203 bits maximum.

21.   The efficiency of the one-dimensional message is 98.5%. For the two-dimensional case, efficiency becomes 98.6% (fewer correctable bits), for the three-dimensional case, 98.5%, and the four-dimensional case is 98.6%.

16.4 CONVOLUTIONAL CODING

1.

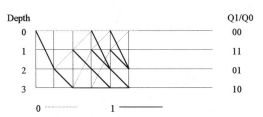

The coding rate is 1/2 or two bits out for each bit in.

2.   The message 01001011 will produce an output stream

$$Q1/Q0 = 00\ 01\ 11\ 00\ 01\ 11\ 01\ 10$$

3.   A message arrives as 00 01 00 10 11 10 11 00 01. We need to generate a cost table for it. Hint: It is quite helpful to preload the cost table with individual choice costs, before attempting to calculate the path costs. Such choice costs are included (small font) in the table below.

Cost	00	01	00	10	11	10	11	00	01
00	0	1	0	1	2	1	2	0	1
	0	1	1/2	2/3	4/4	5/2	4/5	4/2	3/6
11	2	1	2	1	0	1	0	2	1
		2	2/5	3/2	3/1	4/3	3/2	5/5	4/5
01	1	0	1	2	1	2	1	1	0
	1	0	2/3	3/4	3/3	6/3	3/4	5/3	2/5
10	1	2	1	0	1	0	1	1	2
		3	1/4	2/1	4/2	3/2	4/3	4/4	5/6

From the table, the corrected message is 00 01 10 10 10 10 11 00 01 so the data are 011111001.

4.

Cost	10	26	33	62	56	43	77	21	14
00	1	8	6	8	11	6	14	3	5
	1	9	15/20	23/20	31/28	34/23	37/4	40/28	33/46
77	13	6	8	6	3	6	0	11	9
		14	12/27	22/17	26/17	29/28	25/2	41/46	45/45
07	8	3	7	11	6	8	7	8	4
	8	4	16/21	26/23	26/23	36/25	30/3	45/33	32/45
70	6	11	7	3	8	6	7	6	10
		19	11/26	19/14	31/22	29/28	42/3	36/41	46/46

In this case, the message is 00 01 10 10 11 01 11 00 01 so the data are 0 1 1 1 0 1 0 0 1.

5. Since a 1 input always causes an increase in trellis depth (up to 7) the data must enter the three registers at the MSB. A change from state 7 to 3 changes both output bits so D2 goes to both outputs. A change from 3 to 1 again changes both outputs so D1 goes to both. A change from 1 to 0 only changes Q1, so D0 only goes to Q1. This gives the following circuit.

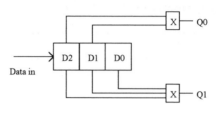

Appendix A

Some primitive polynomials

In this appendix, Tables A.1 to A.25 list some primitive polynomials for a range of field sizes. Because the number of polynomials per field increases rapidly with field size, complete tables are not given for field sizes greater than $GF(2^{12})$. Using the program library listed later, you can calculate more of these for yourself. For convenience, the polynomials are given in base 10.

Table A.1 Primitive polynomials over $GF(2^2)$

Field	Primitive polynomial
$GF(2^2)$	7

Table A.2 Primitive polynomials over $GF(2^3)$

Field	Primitive polynomial
$GF(2^3)$	11,13

Table A.3 Primitive polynomials over $GF(2^4)$

Field	Primitive polynomial
$GF(2^4)$	19,25

Table A.4 Primitive polynomials over $GF(2^5)$

Field	Primitive polynomial
$GF(2^5)$	37,41,47,55,59,61

Table A.5 Primitive polynomials over GF(2^6)

Field	Primitive polynomial
GF(2^6)	67,91,97,103,109,115

Table A.6 Primitive polynomials over GF(2^7)

Field	Primitive polynomial
GF(2^7)	131,137,143,145,157,167,171,185,191,193,203,211,213,229,239,241, 247,253

Table A.7 Primitive polynomials over GF(2^8)

Field	Primitive polynomial
GF(2^8)	285,299,301,333,351,355,357,361,369,391,397,425,451,463,487,501

Table A.8 Primitive polynomials over GF(2^9)

Field	Primitive polynomial
GF(2^9)	529,539,545,557,563,601,607,617,623,631,637,647,661,675,677,687, 695,701,719,721,731,757,761,787,789,799,803,817,827,847,859,865, 875,877,883, 895,901,911,949,953,967,971,973,981,985,995,1001,1019

Table A.9 Primitive polynomials over GF(2^{10})

Field	Primitive polynomial
GF(2^{10})	1033,1051,1063,1069,1125,1135,1153,1163,1221,1239,1255,1267,1279, 1293,1305,1315,1329,1341,1347,1367,1387,1413,1423,1431,1441,1479, 1509,1527,1531,1555,1557,1573,1591,1603,1615,1627,1657,1663,1673, 1717,1729,1747,1759,1789,1815,1821,1825,1849,1863,1869,1877,1881, 1891,1917,1933,1939,1969,2011,2035,2041

Table A.10 Primitive polynomials over GF(2^{11})

Field	Primitive polynomial
GF(2^{11})	2053,2071,2091,2093,2119,2147,2149,2161,2171,2189,2197,2207,2217, 2225,2255,2257,2273,2279,2283,2293,2317,2323,2341,2345,2363,2365, 2373,2377,2385,2395,2419,2421,2431,2435,2447,2475,2477,2489,2503, 2521,2533,2551,2561,2567,2579,2581,2601,2633,2657,2669,2681,2687, 2693,2705,2717,2727,2731,2739,2741,2773,2783,2793,2799,2801,2811, 2819,2825,2833,2867,2879,2881,2891,2905,2911,2917,2927,2941,2951, 2955,2963,2965,2991,2999,3005,3017,3035,3037,3047,3053,3083,3085, 3097,3103,3159,3169,3179,3187,3205,3209,3223,3227,3229,3251,3263, 3271,3277,3283,3285,3299,3305,3319,3331,3343,3357,3367,3373,3393, 3399,3413,3417,3427,3439,3441,3475,3487,3497,3515,3517,3529,3543, 3547,3553,3559,3573,3589,3613,3617,3623,3627,3635,3641,3655,3659, 3669,3679,3697,3707,3709,3713,3731,3743,3747,3771,3791,3805,3827, 3833,3851,3865,3889,3895,3933,3947,3949,3957,3971,3985,3991,3995, 4007,4013,4021,4045,4051,4069,4073

Table A.11 Primitive polynomials over GF(2^{12})

Field	Primitive polynomial
GF(2^{12})	4179,4201,4219,4221,4249,4305,4331,4359,4383,4387,4411,4431,4439, 4449,4459,4485,4531,4569,4575,4621,4663,4669,4711,4723,4735,4793, 4801,4811,4879,4893,4897,4921,4927,4941,4977,5017,5027,5033,5127, 5169,5175,5199,5213,5223,5237,5287,5293,5331,5391,5405,5453,5523, 5573,5591,5597,5611,5641,5703,5717,5721,5797,5821,5909,5913,5955, 5957,6005,6025,6061,6067,6079,6081,6231,6237,6289,6295,6329,6383, 6427,6453,6465,6501,6523,6539,6577,6589,6601,6607,6631,6683,6699, 6707,6761,6795,6865,6881,6901,6923,6931,6943,6999,7057,7079,7103, 7105,7123,7173,7185,7191,7207,7245,7303,7327,7333,7355,7365,7369, 7375,7411,7431,7459,7491,7505,7515,7541,7557,7561,7701,7705,7727, 7749,7761,7783,7795,7823,7907,7953,7963,7975,8049,8089,8123,8125, 8137

Table A.12 Primitive polynomials over GF(2^{13})

Field	Primitive polynomial
GF(2^{13})	8219,8231,8245,8275,8293,8303,8331,8333,8351,8357,8367,8379,8381, 8387,8393,8417,8435,8461,8469,8489,8495,8507,8515,8551,8555,8569, 8585,8599,8605,8639

Table A.13 Primitive polynomials over GF(2^{14})

Field	Primitive polynomial
GF(2^{14})	16427,16441,16467,16479,16507,16553,16559,16571,16573,16591, 16619,16627,16653,16659,16699,16707,16795,16797,16807,16813, 16821,16853, 16857,16881

Table A.14 Primitive polynomials over GF(2^{15})

Field	Primitive polynomial
GF(2^{15})	32771,32785,32791,32813,32821,32863,32887,32897,32903,32915, 32933,32963,32975,32989,32999,33013,33025,33045,33061,33111, 33117,33121, 33133,33157

Table A.15 Primitive polynomials over GF(2^{16})

Field	Primitive polynomial
GF(2^{16})	65581,65593,65599,65619,65725,65751,65839,65853,65871,65885, 65943,65953,65965,65983,65991,66069,66073,66085,66095,66141, 66157,66181, 66193,66209

Table A.16 Primitive polynomials over GF(2^{17})

Field	Primitive polynomial
GF(2^{17})	131081,131087,131105,131117,131123,131135,131137,131157,131177, 131195,131213,131225,131235,131247,131259,131269,131317

Table A.17 Primitive polynomials over GF(2^{18})

Field	Primitive polynomial
GF(2^{18})	262183,262207,262221,262267,262273,262363,262375,262381,262407

Table A.18 Primitive polynomials over GF(2^{19})

Field	Primitive polynomial
GF(2^{19})	524327,524351,524359,524371,524377,524387,524399,524413,524435, 524463

Table A.19 Primitive polynomials over GF(2^{20})

Field	Primitive polynomial
GF(2^{20})	1048585,1048659,1048677,1048681,1048699

Table A.20 Primitive polynomials over GF(2^{21})

Field	Primitive polynomial
GF(2^{21})	2097157,2097191,2097215,2097253

Table A.21 Primitive polynomials over GF(2^{22})

Field	Primitive polynomial
GF(2^{22})	4194307,4194361

Table A.22 Primitive polynomials over GF(2^{23})

Field	Primitive polynomial
GF(2^{23})	8388641,8388651,8388653,8388659,8388671,8388685,8388709

Table A.23 Primitive polynomials over GF(2^{24})

Field	Primitive polynomial
GF(2^{24})	16777243

Table A.24 Primitive polynomials over GF(2^{25})

Field	Primitive polynomial
GF(2^{25})	33554441

Table A.25 Primitive polynomials over GF(2^{26})

Field	Primitive polynomial
GF(2^{26})	67108935

Appendix B

Solutions to some key equations

The probability of making calculation errors when solving the key equations by hand increases rapidly with correctable errors. To provide a quick reference, however, the solutions to a few equations are presented in this appendix for small values of t. These solutions are for frequency domain encoded data, and assume that the recursive extension feedback weights are labelled from right to left, L_1 to L_t, where there are t errors to solve for, and from left to right, the known frequency domain error spectra are E_0 to E_{2t-1}.

$t = 1$:

Key equations

$$E_0.L_1 = E_1$$

so

$$L_1 = E_1/E_0$$

$t = 2$:

Key equations

$$E_0.L_2 + E_1.L_1 = E_2$$

$$E_1.L_2 + E_2.L_1 = E_3$$

so

$$L_1 = (E_0E_3 + E_1E_2)/(E_1E_1 + E_0E_2)$$

$$L_2 = (E_1E_3 + E_2E_2)/(E_1E_1 + E_0E_2)$$

$t = 3$:

Key equations

$$E_0.L_3 + E_1.L_2 + E_2.L_1 = E_3$$

$$E_1.L_3 + E_2.L_2 + E_3.L_1 = E_4$$

$$E_2.L_3 + E_3.L_2 + E_4.L_1 = E_5$$

Rearranging into a matrix gives

$$\begin{bmatrix} E_0 & E_1 & E_2 \\ E_1 & E_2 & E_3 \\ E_2 & E_3 & E_4 \end{bmatrix} \times \begin{bmatrix} L_3 \\ L_2 \\ L_1 \end{bmatrix} = \begin{bmatrix} E_3 \\ E_4 \\ E_5 \end{bmatrix}$$

Inverting the matrix

$$\frac{1}{\Delta} \begin{bmatrix} E_2 E_4 + E_3^2 & E_1 E_4 + E_2 E_3 & E_1 E_3 + E_2^2 \\ E_1 E_4 + E_2 E_3 & E_0 E_4 + E_2^2 & E_0 E_3 + E_1 E_2 \\ E_1 E_3 + E_2^2 & E_0 E_3 + E_1 E_2 & E_0 E_2 + E_1^2 \end{bmatrix} \times \begin{bmatrix} E_3 \\ E_4 \\ E_5 \end{bmatrix} = \begin{bmatrix} L_3 \\ L_2 \\ L_1 \end{bmatrix}$$

where $\Delta = E_0\left(E_2 E_4 + E_3^2\right) + E_1\left(E_1 E_4 + E_2 E_3\right) + E_2\left(E_1 E_3 + E_2^2\right)$

$$L_1 = \left(E_3\left(E_1 E_3 + E_2^2\right) + E_4\left(E_0 E_3 + E_1 E_2\right) + E_5\left(E_0 E_2 + E_1^2\right)\right)\Big/\Delta$$

$$L_2 = \left(E_3\left(E_1 E_4 + E_2 E_3\right) + E_4\left(E_0 E_4 + E_2^2\right) + E_5\left(E_0 E_3 + E_1 E_2\right)\right)\Big/\Delta$$

$$L_3 = \left(E_3\left(E_2 E_4 + E_3^2\right) + E_4\left(E_1 E_4 + E_2 E_3\right) + E_5\left(E_1 E_3 + E_2^2\right)\right)\Big/\Delta$$

$t = 4$:

Writing the key equations as a matrix gives

$$\begin{vmatrix} E_0 & E_1 & E_2 & E_3 \\ E_1 & E_2 & E_3 & E_4 \\ E_2 & E_3 & E_4 & E_5 \\ E_3 & E_4 & E_5 & E_6 \end{vmatrix} \times \begin{vmatrix} L_4 \\ L_3 \\ L_2 \\ L_1 \end{vmatrix} = \begin{vmatrix} E_4 \\ E_5 \\ E_6 \\ E_7 \end{vmatrix}$$

First we must find the determinant of this matrix:

$$\Delta = \begin{pmatrix} E_0\left(E_2\left(E_4E_6 + E_5^2\right) + E_3\left(E_3E_6 + E_4E_5\right) + E_4\left(E_3E_5 + E_4^2\right)\right) + \\ E_1\left(E_1\left(E_4E_6 + E_5^2\right) + E_3\left(E_2E_6 + E_3E_5\right) + E_4\left(E_2E_5 + E_3E_4\right)\right) + \\ E_2\left(E_1\left(E_3E_6 + E_4E_5\right) + E_2\left(E_2E_6 + E_3E_5\right) + E_4\left(E_2E_4 + E_3^2\right)\right) + \\ E_3\left(E_1\left(E_3E_5 + E_4^2\right) + E_2\left(E_2E_5 + E_3E_4\right) + E_3\left(E_2E_4 + E_3^2\right)\right) \end{pmatrix}$$

Creating the adjoint matrix is rather cumbersome, so only the solutions will be presented:

$$L_4 = \begin{pmatrix} E_4\left(E_2\left(E_4E_6 + E_5^2\right) + E_3\left(E_3E_6 + E_4E_5\right) + E_4\left(E_3E_5 + E_4^2\right)\right) + \\ E_5\left(E_1\left(E_4E_6 + E_5^2\right) + E_2\left(E_3E_6 + E_4E_5\right) + E_3\left(E_3E_5 + E_4^2\right)\right) + \\ E_6\left(E_1\left(E_3E_6 + E_4E_5\right) + E_2\left(E_2E_6 + E_4^2\right) + E_3\left(E_2E_5 + E_3E_4\right)\right) + \\ E_7\left(E_1\left(E_3E_5 + E_4^2\right) + E_2\left(E_2E_5 + E_3E_4\right) + E_3\left(E_2E_4 + E_3^2\right)\right) \end{pmatrix} \frac{1}{\Delta}$$

$$L_3 = \begin{pmatrix} E_4\left(E_1\left(E_4E_6 + E_5^2\right) + E_3\left(E_2E_6 + E_3E_5\right) + E_4\left(E_2E_5 + E_3E_4\right)\right) + \\ E_5\left(E_0\left(E_4E_6 + E_5^2\right) + E_2\left(E_2E_6 + E_3E_5\right) + E_3\left(E_2E_5 + E_3E_4\right)\right) + \\ E_6\left(E_0\left(E_3E_6 + E_4E_5\right) + E_2\left(E_1E_6 + E_3E_4\right) + E_3\left(E_1E_5 + E_3^2\right)\right) + \\ E_7\left(E_0\left(E_3E_5 + E_4^2\right) + E_2\left(E_1E_5 + E_2E_4\right) + E_3\left(E_1E_4 + E_2E_3\right)\right) \end{pmatrix} \frac{1}{\Delta}$$

$$L_2 = \begin{pmatrix} E_4\left(E_1\left(E_3E_6 + E_4E_5\right) + E_2\left(E_2E_6 + E_3E_5\right) + E_4\left(E_2E_4 + E_3^2\right)\right) + \\ E_5\left(E_0\left(E_3E_6 + E_4E_5\right) + E_1\left(E_2E_6 + E_3E_5\right) + E_3\left(E_2E_4 + E_3^2\right)\right) + \\ E_6\left(E_0\left(E_2E_6 + E_4^2\right) + E_1\left(E_1E_6 + E_3E_4\right) + E_3\left(E_1E_4 + E_2E_3\right)\right) + \\ E_7\left(E_0\left(E_2E_5 + E_3E_4\right) + E_1\left(E_1E_5 + E_2E_4\right) + E_3\left(E_1E_3 + E_2^2\right)\right) \end{pmatrix} \frac{1}{\Delta}$$

$$L_1 = \begin{pmatrix} E_4\left(E_1\left(E_3E_5 + E_4^2\right) + E_2\left(E_2E_5 + E_3E_4\right) + E_3\left(E_2E_4 + E_3^2\right)\right) + \\ E_5\left(E_0\left(E_3E_5 + E_4^2\right) + E_1\left(E_2E_5 + E_3E_4\right) + E_2\left(E_2E_4 + E_3^2\right)\right) + \\ E_6\left(E_0\left(E_2E_5 + E_3E_4\right) + E_1\left(E_1E_5 + E_3^2\right) + E_2\left(E_1E_4 + E_2E_3\right)\right) + \\ E_7\left(E_0\left(E_2E_4 + E_3^2\right) + E_1\left(E_1E_4 + E_2E_3\right) + E_2\left(E_1E_3 + E_2^2\right)\right) \end{pmatrix} \frac{1}{\Delta}$$

Appendix C

Frequency domain example

This appendix gives Pascal program listings for three frequency domain error correction examples. The listings start with a unit or library of functions used by all subsequent program examples. The library, GFUn (or Galois Field Unit) contains the essential field element manipulation and display procedures and functions which underpin the examples. Two other libraries, GFMatUn and SKeyUn, required for the second and third examples may be found on the Internet. The listings for these latter libraries are included here.

The three programming examples start with a hand-solved triple-error problem over $GF(2^8)$. This is followed by two eight-error problems using the GFMatUn and SKeyUn libraries respectively. Some of the functions are written in Assembly, but for the more common of these the Pascal equivalent is also included. This typically gives a tenfold speed-up in program execution. The library listed here is designed to work on field sizes up to and including $GF(2^8)$. All examples were written using Borland's Turbo Pascal® 7.0. Turbo Pascal is a registered trademark of Borland International, Inc.

Unit GFUn;
Interface

Type	BA =	Array [0..255] Of Byte;	
	BP =	^BA;	
Function	**GFGen**	(Key : Word)	: Boolean;
Function	**GF2Real**	(a : Byte)	: Byte;
Function	**Real2GF**	(a : Byte)	: Byte;
Function	**AaB**	(a, b : Byte)	: Byte;
Function	**AxB**	(a, b : Byte)	: Byte;
Function	**AoB**	(a, b : Byte)	: Byte;
Function	**ApB**	(a, b : LongInt)	: Byte;
Function	**ArB**	(a, b: LongInt)	: Byte;
Function	**PrimPoly**	(Field, Start : LongInt)	: LongInt;
Function	**Factors**	(Poly, Start : LongInt; Var F1, F2 : LongInt)	: Boolean;
Procedure	**TimesArray**	(D : BP; b : Byte);	
Procedure	**DivideArray**	(D : BP; b : Byte);	
Procedure	**IFT**	(Pin, Pout : BP);	
Procedure	**IFT2**	(PIn, POut : BP);	
Procedure	**FFT2**	(PIn, POut : BP);	
Procedure	**DisplayDec**	(D : BP; s, x, y : Byte);	
Procedure	**DisplayChar**	(D : BP; s, x, y, c : Byte);	
Procedure	**DisplayHex**	(D : BP; s, x, y : Byte);	
Procedure	**DisplayHexPower**	(D : BP; s, x, y : Byte);	
Procedure	**DisplayDecPower**	(D : BP; s, x, y : Byte);	
Procedure	**Display2D**	(P : BP; x, y : Byte);	
Function	**GetGFPointer**		: Pointer;
Function	**GetGIPointer**		: Pointer;
Function	**GetGFSize**		: Byte;
Function	**NZ**	(a : LongInt)	: Byte;
Procedure	**ListPrims**	(a : LongInt);	
Procedure	**ListFactors**	(P : LongInt);	
Function	**GetPair**	(a : LongInt)	: LongInt;

{**Designed to calculate/manipulate any field up to GF(2^8) given a key.**}
Implementation
Uses CRT;

Var	GF, GI	: BA;
	FL	: Byte;
	PF, PI	: Pointer;

{**GF turns powers to bit patterns, GI turns bit patterns to powers. These functions allow programs to gain access to the field element look-up tables and size.**}

Function	**GetGFPointer**	: Pointer;

Begin

    GetGFPointer := @GF;

End;

Function	**GetGIPointer**	: Pointer;

Begin

    GetGIPointer := @GI;

End;

Function	**GetGFSize**	: Byte;

Begin

    GetGFSize := FL;

End;

{The following two procedures calculate the forward and reverse look-up tables of elements, using a primitive polynomial supplied to GFGen. GFGen returns the value true if it was able to generate the field (i.e. the supplied key was primitive).}

```
Procedure   GIGen;
Var         i                                                    : Word;
Begin
            For i := 0 to FL do GI[GF[i]] := i;
End;
Function    GFGen (Key : Word)                                   : Boolean;
Var         i, a, b                                              : Word;
Begin
            FL := (1 SHL Trunc (Ln (Key)/Ln(2)))-1;
            GFGen := True;
            GF [0] := 0;
```

{Strictly, GF[0] should be 1, i.e. α^0, but we must have 0 since it is a member of the field. This programming convenience must be remembered.}

```
            a := 2; b := FL + 1;
            For i := 1 to FL do
            Begin
                GF [i] := a;
                a := a shl 1;
                If (a and b) <> 0 Then a := a XOR Key;
                If (a=2) AND (i<FL) Then GFGen := False;
            End;
            GIGen;
End;
```

{The following two functions allow movement between pattern and power representations of field elements using the GF and GI look-up tables.}

```
Function    GF2Real (a : Byte)                                   : Byte;
Begin
            GF2Real := GF [a];
End;
Function    Real2GF (a : Byte)                                   : Byte;
Begin
            Real2GF := GI [a];
End;
```

{InRange is an internal function which returns any input value into a power of α. If 0 is input, the result will be returned as FL, the maximum power in the field since $\alpha^0 = \alpha^{FL}$. NZ is identical to InRange but is declared publicly. This allows programs to treat the 0 element as 1 so, for example, calculating part of a Fourier transform using

$$AxB\ (d[i],\ NZ(i*j))$$

would produce the correct result whereas

$$AxB\ (d[i],\ i*j)$$

would not since if either *i* or *j* are zero, their meaning will be misinterpreted.}

```
Function    InRange (a : LongInt)                                : Byte;
```

Begin

       If a < 0   Then a := FL - Abs (a) MOD FL Else

       If a > FL   Then a := a MOD FL;

       If a = 0   Then a := FL;

       InRange := a;

End;

Function	**NZ** (a : LongInt)	: Byte;

Begin

       NZ := InRange (a);

End;

{**AaB adds elements, AxB multiplies, AoB divides, ApB raises to power and ArB lowers to root. All inputs are treated as powers although 0 is treated as the zero member of the field.**}

Function	**AaB** (a, b : Byte)	: Byte;

Begin

       Asm

          les si, [PF]

          xor bh, bh

          mov bl, a

          mov al, [es:si+bx]

          mov bl, b

          xor al, [es:si+bx]

          mov bl, al

          les si, [PI]

          mov al, [es:si+bx]

          mov @result, al

       End;

End;

Function	**AxB** (a, b : Byte)	: Byte;
Var	i	: Word;

Begin

       i := FL;

       Asm

          mov al, a

          cmp al, 0

          jz @End

          mov bl, al

          mov al, b

          cmp al, 0

          jz @End

          mov dx, i

          xor ah, ah

          xor bh, bh

          add ax, bx

          cmp ax, dx

          jle @End

          sub ax, dx

          @End:

          mov @result, al

       End;

End;

Function	**AaB1** (a, b : Byte)	: Byte;
Begin		
	AaB1 := GI[GF [a] XOR GF[b]];	
End;		

Function	**AxB1** (a, b : Byte)	: Byte;
Var	i	: Integer;
Begin		
	If (a<>0) AND (b<>0) Then	
	Begin	
	i := a + b;	
	AxB1 := InRange (i);	
	End Else AxB1 := 0;	
End;		

Function	**AoB** (a, b : Byte)	: Byte;
Var	i	: Word;
Begin		
	i := FL;	
	Asm	
	mov al, b	
	cmp al, 0	
	jz @End	
	mov bl, al	
	mov al, a	
	cmp al, 0	
	jz @End	
	xor ah, ah	
	xor bh, bh	
	cmp ax, bx	
	jg @Next	
	add ax, i	
	@Next:	
	sub ax, bx	
	@End:	
	mov @result, al	
	End;	
End;		

Function	**AoB1** (a, b : Byte)	: Byte;
Var	i	: Integer;
Begin		
	If (a<>0) AND (b<>0) Then	
	Begin	
	i := a;	
	i := i - b;	
	AoB1 := InRange (i);	
	End Else AoB1 := 0;	
End;		

Function	**ApB** (a, b : LongInt)	: Byte;
Var	i	: LongInt;
Begin		
	i := a * b;	
	ApB := InRange (i);	
End;		

```
Function    ArB (a, b : LongInt)                                              : Byte;
Var         i                                                                 : Integer;
Begin
            If (b>0) AND (a>0) Then
            Begin
                i := 0;
                While (i<FL) AND (a Mod b<>0) do
                Begin
                    a := a+FL;
                    Inc (i);
                End;
                If i<FL Then ArB := InRange (a DIV b) Else ArB := 0;
            End Else ArB := 0;
End;
```

{**The following multiply or divide all the contents of an array by a constant value.**}

```
Procedure   TimesArray (D : BP; b : Byte);
Var         i                                                                 : Word;
Begin
            For i := 0 to FL-1 do D^ [i] := AxB(D^ [i], b);
End;
```

```
Procedure   DivideArray (D : BP; b : Byte);
Var         i                                                                 : Word;
Begin
            For i := 0 to FL-1 do D^ [i] := AoB(D^ [i], b);
End;
```

{**Performs a forward and inverse Fourier transform, assumes patterns not powers in array pointed to by Pin.**}

```
Procedure   FFT1 (Pin, Pout : BP);
Var         i, j                                                              : LongInt;
            a                                                                 : Byte;
Begin
            For i := 0 to FL-1 do
            Begin
                a := 0;
                For j := 0 to FL-1 do a := a XOR GF[AxB (ApB (1, i*j), GI[Pin^[j]])];
                Pout^[i] := a;
            End;
End;
```

```
Procedure   IFT1 (Pin, Pout : BP);
Var         i, j                                                              : LongInt;
            a                                                                 : Byte;
Begin
            For i := 0 to FL-1 do
            Begin
                a := 0;
                For j := 0 to FL-1 do a := a XOR GF[AxB (ApB (1, -i*j), GI[Pin^[j]])];
                Pout^[i] := a;
            End;
End;
```

```
Procedure   FFT(Pin, Pout : BP);
Var         i, j, k                                        : Word;
            a                                              : Byte;
            P1,P2,PF,PI                                    : Pointer;
Begin
            k := FL;
            P1 := PIn;
            P2 := POut;
            PF := @GF;
            PI := @GI;
            For i := 0 to FL-1 do
            Begin
                a := 0;
                Asm
                    mov cx, k
                    les si, [P1]
                    push ds
                    xor ax, ax
                    xor bx, bx
                    mov dx, k
                    @LOOP:
                    lds di, [PI]
                    mov bl, [es:si]
                    inc si
                    cmp bl, 0
                    jz @END3
                    mov al, [ds:di+bx]
                    add ax, dx
                    cmp ax, k
                    jle @END1
                    sub ax, k
                    @END1:
                    mov bl, al
                    lds di, [PF]
                    mov al, a
                    xor al, [ds:di+bx]
                    mov a, al
                    @END3:
                    add dx, i
                    cmp dx, k
                    jle @END2
                    sub dx, k
                    @END2:
                    loop @LOOP
                    pop ds
                end;
                Pout^[i] := a;
            End;
End;
```

```
Procedure    IFT (Pin, Pout : BP);
Var          i, j, k                              : Word;
             a                                    : Byte;
             P1,P2,PF,PI                          : Pointer;
Begin
             k := FL;
             P1 := PIn;
             P2 := POut;
             PF := @GF;
             PI := @GI;
             For i := 0 to FL-1 do
             Begin
                 a := 0;
                 Asm
                     mov cx, k
                     les si, [P1]
                     push ds
                     xor ax, ax
                     xor bx, bx
                     mov dx, k
                     @LOOP:
                     lds di, [PI]
                     mov bl, [es:si]
                     inc si
                     cmp bl, 0
                     jz @END3
                     mov al, [ds:di+bx]
                     add ax, dx
                     cmp ax, k
                     jle @END1
                     sub ax, k
                     @END1:
                     mov bl, al
                     lds di, [PF]
                     mov al, a
                     xor al, [ds:di+bx]
                     mov a, al
                     @END3:
                     add dx, k
                     sub dx, i
                     cmp dx, k
                     jle @END2
                     sub dx, k
                     @END2:
                     loop @LOOP
                     pop ds
                 end;
                 Pout^[i] := a;
             End;
End;
```

{The following are two-dimensional forward and inverse transforms. They assume the data set is
the field size squared. No Assembly version is given.}

```
Procedure    FFT2 (PIn, POut : BP);
Var          i, j, x, y                                              : Integer;
             l                                                       : Byte;
Begin
             For i := 0 to FL-1 do
             For j := 0 to FL-1 do
             Begin
                 l := 0;
                 For x := 0 to FL-1 do
                 For y := 0 to FL-1 do
                 l := l XOR GF2Real (AxB (Real2GF (Pin^[x*Fl+y]), ApB (1,(i+y)*(j+x))));
                 POut^[i*FL+j] := l;
             End;
End;
```

```
Procedure    IFT2 (PIn, POut : BP);
Var          i, j, x, y                                              : Integer;
             l                                                       : Byte;
Begin
             For i := 0 to FL-1 do
             For j := 0 to FL-1 do
             Begin
                 l := 0;
                 For x := 0 to FL-1 do
                 For y := 0 to FL-1 do
                 l := l XOR GF2Real (AxB (Real2GF (Pin^[x*Fl+y]), ApB (1,-(i+y)*(j+x))));
                 POut^[i*FL+j] := l;
             End;
End;
```

{Sundry display utilities.}

```
Procedure    DisplayDec (D : BP; s, x, y : Byte);
Var          i, j                                                    : Byte;
Begin
             TextColor (14);
             For i := 0 to 15 do
             Begin
                 GotoXY (x, y+i);
                 For j := 0 to 15 do
                 Begin
                     If (i AND j <> 15) AND (j+i SHL 4<= s) Then
                     Write (D^[i shl 4 or j]:4);
                 End;
             End;
End;
```

```
Procedure   DisplayChar  (D : BP; s, x, y, c : Byte);
Var         i, j                                                    : Byte;
Begin
            TextColor (c);
            For i := 0 to 15 do
            Begin
                GotoXY (x, y+i);
                For j := 0 to 15 do
                Begin
                    If (i AND j <> 15) AND (j+i SHL 4<= s) Then
                    Case D^[i shl 4 or j]of
20..254:                Write (Chr (D^[i shl 4 or j]));
0..255:                 Write ('.');
                    End;
                End;
            End;
End;
```
```
Procedure   DisplayHex (D : BP; s, x, y : Byte);
Var         i, j                                                    : Byte;
Begin
            TextColor (14);
            For i := 0 to 15 do
            Begin
                GotoXY (x, i+y);
                For j := 0 to 15 do
                Begin
                    If (i AND j <> 15) AND (j+i SHL 4<= s) Then
                    Begin
                        Write (Copy ('0123456789ABCDEF',1+D^[i shl 4 or j]SHR 4, 1));
                        Write (Copy ('0123456789ABCDEF',1+D^[i shl 4 or j]AND 15, 1),' ');
                    End;
                End;
            End;
End;
```
```
Procedure   DisplayHexPower(D : BP; s, x, y : Byte);
Var         i, j                                                    : Byte;
Begin
            TextColor (14);
            For i := 0 to 15 do
            Begin
                GotoXY (x, i+y);
                For j := 0 to 15 do
                Begin
                    If (i AND j <> 15) AND (j+i SHL 4<= s) Then
                    Begin
                        Write (Copy ('0123456789ABCDEF',1+GI[D^[i shl 4 or j]]SHR 4, 1));
                        Write (Copy ('0123456789ABCDEF',1+GI[D^[i Shl 4 or j]]AND 15, 1),' ');
                    End;
                End;
            End;
End;
```

```
Procedure   DisplayDecPower (D : BP; s, x, y : Byte);
Var         i, j                                                    : Byte;
Begin
            TextColor (14);
            For i := 0 to 15 do
            Begin
                GotoXY (x, i+y);
                For j := 0 to 15 do
                Begin
                    If (i AND j <> 15) AND (j+i SHL 4<= s) Then
                    Begin
                        Write (GI[D^[i shl 4 or j]]:4);
                    End;
                End;
            End;
End;

Procedure   Display2D (P : BP; x, y : Byte);
Var         i, j                                                    : Integer;
Begin
            For j := 0 to FL-1 do
            Begin
                GotoXY (x, Y+j);
                For i := 0 to FL-1 do
                Begin
                    If P^[i*FL+j]=0 Then TextColor (10) Else TextColor (7);
                    Write (P^[i*FL+j]:2);
                End;
            End;
End;
```

{**Designed to calculate primitive polynomials.**}

```
Function    PrimPoly (Field, Start : LongInt)                       : LongInt;
Var         i, j, a, Len                                            : LongInt;
Begin
            Inc (Start);                    Len := (1 SHL Field);
            Start := Start OR (Len OR 1);   PrimPoly := 0;
            If (Field > 1) Then
            Begin
                If Start<(Len shl 1) - 1 Then
                For j := Start to (Len shl 1) - 1 do
                Begin
                    a := 1;i := 0;
                    If j and 1 = 1 Then
                    Repeat
                        Inc (i);a := a shl 1;
                        If (a and Len) <> 0 Then a := a XOR j;
                    Until (a = 1) OR (i = Len);
                    If (i = Len - 1) Then
                    Begin
                        PrimPoly := j; j := (Len shl 1) - 1;
                    End;
                End;
            End;
End;
```

{This function attempts to factorize a polynomial. It returns the factors in **F1** and **F2**. Since there may be more than two factors, the routine can be preloaded via start to pick out successive ones.}

```
Function    Factors (Poly, Start : LongInt; Var F1, F2 : LongInt)        : Boolean;
Var         n, i, j, k, l, m, p, q, R                                     : LongInt;
Begin
            Factors := False;
            Inc (Start);
            If Start<3 Then Start := 3;
            m := Round (Ln (Poly)/Ln (2));
            If (Start<Poly) Then
            For i := Start to Poly do
            If i AND 1=1 Then
            Begin
                n := Round (Ln (i)/Ln (2));
                For j := 3 to Poly do
                If j AND 1=1 Then
                Begin
                    q := i; p := j; R := 0;
                    For k := 0 to n do
                    Begin
                        If q AND 1 = 1 Then R := R XOR (p);
                        q := q SHR 1;
                        p := p SHL 1;
                    End;
                    If R>Poly Then j := Poly Else
                    If R=Poly Then
                    Begin
                        Factors := True;
                        F1 := i; F2 := j;
                        j := Poly; i := Poly;
                    End;
                End;
            End;
End;
```

{Calculates paired polynomial.}

```
Function    GetPair (a : LongInt)                                         : LongInt;
Var         i        : LongInt;
Begin
            i := 0;
            While (a>0) Do
            Begin
                i := (i SHL 1) OR (a AND 1);
                a := a SHR 1;
            End;
            GetPair := i;
End;
```

{List out primitive polynomials starting at field a. Also show paired polynomial.}

```
Procedure    ListPrims (a : LongInt);
Var          i, j                                                    : LongInt;
Begin
             ClrScr; i := a;
             Repeat
                 TextColor (10);WriteLn ('Field : ',i);
                 j := 0;
                 Repeat
                     j := PrimPoly (i, j);
                     If j<>0 Then
                     Begin
                         TextColor (14); Write (j:8);
                         TextColor (14); Write (GetPair (j):8);
                     End;
                 Until (j=0) OR KeyPressed;
                 WriteLn;
                 Inc (i);
             Until KeyPressed;
End;
```

{List out factors of polynomial P.}

```
Procedure    ListFactors (P : LongInt);
Var          F1, F2                                                  : LongInt;
Begin
             F1 := 0;
             While Factors (P, F1, F1, F2) Do WriteLn (F1, ' ', F2);
End;
```

{Sets up initial field to GF(2^8) using 285.}

```
Begin
             GFGen (285); {Useful field over 255}
             PF := @GF;
             PI := @GI;
End.
```

{In this first programming example, only the library GFUn is used. Solving of the key equations and subsequent recursive extension is performed longhand. The program corrects three errors in a 255-symbol message over GF(2^8).}

```
Program CorrectErrors;
Uses CRT, GFUn;
```

{Primitive polynomials for GF(8) are
285,299,301,333,351,355,357,361,369,397,425,451,463,487,501}

Var		
	TextIn : BA; {Source text file, picked up from disk}	
	TextTD : BA; {Text plus zeros in time domain}	
	TextEr : BA; {Text in time domain plus errors}	
	TextRx : BA; {Text plus errors back in frequency domain}	
	ErrSpc : BA; {Complete error spectrum in frequency domain}	
	ErrPat : BA; {Time domain error pattern}	
	TextEC : BA; {Error corrected text in time domain}	
	TextOu : BA; {Original text after error recovery}	
	Locate : BA; {Error locator vector, calculated for fun}	
{Six known error spectra}		
	E0, E1, E2, E3, E4, E5	: Byte;
{Registers used in REC. Q0 is an intermediate variable}		
	Q0, Q1, Q2, Q3	: Byte;
{Locator feedback taps}		
	L1, L2, L3	: Byte;
{Adj Matrix, used to solve L1-3}		
	A00, A01, A02,	
	A10, A11, A12,	
	A20, A21, A22	: Byte;
{Matrix determinant plus sundry variables}		
	Det, c	: Byte;
	i, j	: Word;
	Ftn	: File;

```
Begin
        ClrScr;
{Generate the finite field}
        If GFGen (285) Then
        Begin
{Pre-load 16 zeros}
                FillChar (TextIn, 16, 0);
{Load in source text data}
        Assign (Ftn, 'test.txt');
                Reset (Ftn, 1);
                BlockRead (Ftn, TextIn[16], 255-16);
                Close (Ftn);
{Display initial data}
                DisplayChar (@TextIn, 255, 1, 3, 13);
                TextColor (10); GotoXY (2, 1);Write ('Original  data');
{Convert to time domain}
                IFT (@TextIn, @TextTD);
{Display message}
                DisplayChar (@TextTD, 255, 19, 3, 11);
                TextColor (10);GotoXY (20, 1);Write ('Encoded message');
```

```
{Copy message to receiver}
          Move (TextTD, TextEr, 255);
```

```
{Add 3 errors to message}
          TextEr [$46] := TextEr [$46] XOR $61;
          TextEr [$78] := TextEr [$78] XOR $84;
          TextEr [$E1] := TextEr [$E1] XOR $18;
```

```
{Display corrupted message}
          DisplayChar (@TextEr, 255, 37, 3, 11);
          TextColor (10);GotoXY (37, 1);Write ('Corrupted message');
```

```
{Generate FD message at Rx}
          FFT (@TextEr, @TextRx);
```

```
{Display corrupted message}
          DisplayChar (@TextRx,255, 55, 3, 13);
          TextColor (10);GotoXY (56, 1);Write ('Corrupted data');
```

```
{Display known error spectrum}
          DisplayHex (@TextRX, 15, 12, 22);
          TextColor (10);GotoXY (13, 20);
          Write ('Top row of Error spectrum from Corrupted data');
```

```
{Pick off first six error spectra since we know that there are only three errors. Also convert data
patterns to powers for solving}

          E0 := Real2GF (TextRx[0]);
          E1 := Real2GF (TextRx[1]);
          E2 := Real2GF (TextRx[2]);
          E3 := Real2GF (TextRx[3]);
          E4 := Real2GF (TextRx[4]);
          E5 := Real2GF (TextRx[5]);
```

```
{Calculate determinant longhand}

          Det :=AaB (AxB (E0, AaB (AxB (E2, E4), AxB(E3, E3))),
                 AaB (AxB (E1, AaB (AxB (E1, E4), AxB(E2, E3))),
                 AxB (E2, AaB (AxB (E1, E3), AxB(E2, E2)))));
```

```
{Calculate Adjoint matrix. Note also shared terms}

          A00 := AaB (AxB (E2, E4), AxB (E3, E3));
          A01 := AaB (AxB (E1, E4), AxB (E2, E3));
          A02 := AaB (AxB (E1, E3), AxB (E2, E2));

          A10 := A01;
          A11 := AaB (AxB (E0, E4), AxB (E2, E2));
          A12 := AaB (AxB (E0, E3), AxB (E1, E2));

          A20 := A02;
          A21 := A12;
          A22 := AaB (AxB (E0, E2), AxB (E1, E1));
```

{Solve matrix for L1-L3}
```
        L3 := AoB (AaB (AaB (AxB (A00, E3), AxB (A01, E4)),
                            AxB (A02, E5)), Det);
        L2 := AoB (AaB (AaB (AxB (A10, E3), AxB (A11, E4)),
                            AxB (A12, E5)), Det);
        L1 := AoB (AaB (AaB (AxB (A20, E3), AxB (A21, E4)),
                            AxB (A22, E5)), Det);
```

{Pre-load REC with first three known error spectra}
```
        Q3 := E0;
        Q2 := E1;
        Q1 := E2;
```

{Complete error spectrum. Calculate feedback in Q0}
```
        For i := 0 to 254 do
        Begin
                ErrSpc [i] := GF2Real (Q3);
                Q0 := AaB (AaB(        AxB(Q3, L3),
                                       AxB (Q2, L2)),
                                       AxB (Q1, L1));
```

{Shift registers}
```
                Q3 := Q2; Q2 := Q1; Q1 := Q0;
        End;
```

{Display completed spectrum}
```
        DisplayChar (@ErrSpc, 255, 1, 26, 13);
        TextColor (10);GotoXY (2, 24);Write ('Error spectrum');
```

{Construct time domain errors}
```
        IFT (@ErrSpc, @ErrPat);
```

{Display errors}
```
        DisplayChar (@ErrPat, 255, 19, 26, 11);
        TextColor (10);GotoXY (19, 24);Write ('TD Error Pattern');
```

{Correct time domain message}
```
        For i := 0 to 254 do
                TextEC [i] := TextEr [i] XOR ErrPat[i];
```

{Convert corrected message to frequency domain and display}
```
        FFT (@TextEC, @TextOu);
        DisplayChar (@TextOu, 255, 37, 26, 13);
        TextColor (10);GotoXY (38, 24);Write ('Corrected Data');
```

{Generate error locator for fun}
```
        For i := 0 to 254 do
                Locate [i] := GF2Real (    AaB (AaB (NZ(3*i), NZ(2*i+L1)),
                                            AaB (NZ(i+L2),NZ(L3))));
```

{Convert 0 to ASCII 0 in order to view}
```
        For i := 0 to 254 do If Locate [i]=0 Then Locate[i] := $30;
```

{Display error locator}
```
        DisplayChar (@Locate, 255, 55, 26, 11);
        TextColor (10);GotoXY (57, 24);Write ('Error locator');
```

```
{Display information}
            GotoXY (12, 48);TextColor (9);
            Write ('FREQUENCY DOMAIN CODING EXAMPLE WITH 3 ERRORS');
            TextColor (13); GotoXY (10, 45);
            Write ('Magenta = FD');
            TextColor (11); GotoXY (30, 45);
            Write ('Cyan   = TD');
            TextColor (14); GotoXY (50, 45);
            Write ('Yellow  = FD/HEX');
      End;
      ReadKey;
End.
```

{The next example corrects eight errors, but uses the matrix unit to simplify the inversion. The chances of not making a mistake while trying to invert an 8x8 matrix longhand are very small.}

```
Program Correct8Errors;
Uses CRT, GFMatUn, GFUn;
Var    TextIn      : BA; {Source text file, picked up from disk}
       TextTD      : BA; {Text plus zeros in time domain}
       TextEr      : BA; {Text in time domain plus errors}
       TextRx      : BA; {Text plus errors back in frequency domain}
       ErrSpc      : BA; {Complete error spectrum in frequency domain}
       ErrPat      : BA; {Time domain error pattern}
       TextEC      : BA; {Error corrected text in time domain}
       TextOu      : BA; {Original text after error recovery}
       Locate      : BA; {Error locator vector, calculated for fun}

       E           : Array [0..15] Of Byte;   {15 known error spectra}
       Q           : Array [0..8] Of Byte;    {Registers used in REC}
       L           : Array [1..8] Of Byte;    {Locator feedback taps}

       ME,MI,ML    : MatP;                     {Matrices used in solution}

       i, j        : Word;
       Ftn         : File;
Begin
       ClrScr;
{Generate the finite field}
       If GFGen (285) Then
       Begin
{Pre-load 16 zeros}
             FillChar (TextIn, 16, 0);
{Load in source text data}
             Assign (Ftn, 'test.txt');
             Reset (Ftn, 1);
             BlockRead (Ftn, TextIn[16], 255-16);
             Close (Ftn);
```

{Display initial (frequency domain) text}
DisplayChar (@TextIn, 255, 1, 3, 13);
TextColor (10);GotoXY (2, 1);Write ('Original data');

{Convert to time domain}
IFT (@TextIn, @TextTD);

{Display time domain message}
DisplayChar (@TextTD, 255, 19, 3, 11);
TextColor (10);GotoXY (20, 1);Write ('Encoded message');

{Copy message to receiver}
Move (TextTD, TextEr, 255);

{Add eight errors to message}
TextEr [$46]   := TextEr [$46] XOR $61;
TextEr [$78]   := TextEr [$78] XOR $84;
TextEr [$E1]   := TextEr [$E1] XOR $18;
TextEr [$22]   := TextEr [$22] XOR $AA;
TextEr [$4F]   := TextEr [$4F] XOR $D2;
TextEr [$A4]   := TextEr [$A4] XOR $C3;
TextEr [$CC]   := TextEr [$CC] XOR $30;
TextEr [$34]   := TextEr [$34] XOR $23;

{Display corrupted message}
DisplayChar (@TextEr, 255, 37, 3, 11);
TextColor (10);GotoXY (37, 1);Write ('Corrupted message');

{Generate FD message at Rx}
FFT (@TextEr, @TextRx);

{Display corrupted data}
DisplayChar (@TextRx,255, 55, 3, 13);
TextColor (10);GotoXY (56, 1);Write ('Corrupted data');

{Display known error spectrum in Hex}
DisplayHex (@TextRX, 15, 12, 22);
TextColor (10);GotoXY (13, 20);
Write ('Top row of Error spectrum from Corrupted data');
For i := 0 to 15 do E[i]:=TextRx[i];   -

{Pick off 16 error spectra and arrange into an 8x8 matrix

```
    E0,E1,E2,E3,E4,E5,E6,E7          L8      E8
    E1,E2,E3,E4,E5,E6,E7,E8          L7      E9
    E2,E3,E4,E5,E6,E7,E8,E9          L6      EA
    E3,E4,E5,E6,E7,E8,E9,EA X        L5  =   EB
    E4,E5,E6,E7,E8,E9,EA,EB          L4      EC
    E5,E6,E7,E8,E9,EA,EB,EC          L3      ED
    E6,E7,E8,E9,EA,EB,EC,ED          L2      EE
    E7,E8,E9,EA,EB,EC,ED,EE          L1      EF
}
```

{Create 8x8 matrix}
ME := MakeSMatrix (8);

{Load in error spectra}
For i := 0 to 7 do Move (E[i], ME^.P^[8*i], 8);

{Invert the matrix}
MI := InvertMatrix (ME);

{Free memory used by ME}
KillSMatrix (ME);

{Create a 1x8 matrix}

```
                  ME := MakeRMatrix (1, 8);
```

{Copy R/H errors into ME}
```
                  Move (E[8], ME^.P^[0], 8);
```

{X inverse matrix by ME}
```
                  ML := TimesMatrix (MI, ME);
```

{Free-up matrix memory}
```
                  KillSMatrix (MI);
                  KillRMatrix (ME);
```

{Copy solution to L}
```
                  For i := 0 to 7 do
                          L[8-i]:=Real2GF(ML^.P^[i]);
```

{Free-up memory}
```
                  KillRMatrix (ML);
```

{Pre-load REC in Q}
```
                  For i := 0 to 7 do Q[8-i]:=Real2GF(E[i]);
```

{Complete error spectrum, Q[0] used for feedback}
```
                  For i := 0 to 254 do
                  Begin
                          ErrSpc [i] := GF2Real (Q[8]);
                          Q[0] := 0;
                          For j := 8 Downto 1 do
                          Begin
                                  Q[0] := AaB (Q[0], AxB (Q[j], L[j]));
                                  Q[j] := Q[j-1];
                          End;
                  End;
```

{Display completed spectrum and pattern}
```
                  DisplayChar (@ErrSpc, 255, 1, 26, 13);
                  TextColor (10);GotoXY (2, 24);Write ('Error spectrum');

                  IFT (@ErrSpc, @ErrPat);
                  DisplayChar (@ErrPat, 255, 19, 26, 11);
                  TextColor (10);GotoXY (19, 24);Write ('TD Error Pattern');
```

{Correct time domain message}
```
                  For i := 0 to 254 do
                          TextEC [i] := TextEr [i] XOR ErrPat[i];
```

{Convert time domain message back to frequency domain data}
```
                  FFT (@TextEC, @TextOu);
                  DisplayChar (@TextOu, 255, 37, 26, 13);
                  TextColor (10);GotoXY (38, 24);Write ('Corrected Data');
```

{Generate error locator for fun}
```
                  For i := 0 to 254 do
                  Begin
                          Locate [i] := NZ(8*i);
                          For j := 1 to 8 do
                                  Locate [i] := AaB(Locate [i], NZ((8-j)*i+L[j]));
                          Locate [i] := GF2Real (Locate[i]);
                  End;
```

{Convert 0 to ASCII 0}
```
                  For i := 0 to 254 do If Locate [i]=0 Then Locate[i] := $30;
                  DisplayChar (@Locate, 255, 56, 26, 11);
                  TextColor (10);GotoXY (57, 24);Write ('Error locator');
```

{Display information}
```
            GotoXY (12, 48);TextColor (9);
            Write ('FREQUENCY DOMAIN CODING EXAMPLE WITH 8 ERRORS');
            TextColor (13); GotoXY (10, 45);
            Write ('Magenta = FD');
            TextColor (11); GotoXY (30, 45);
            Write ('Cyan    = TD');
            TextColor (14); GotoXY (50, 45);
            Write ('Yellow  = FD/HEX');
      End;
      Repeat Until KeyPressed;
End.
```

{The final example program of this appendix again corrects eight errors, but uses the SKeyUn to speed up and simplify the operation.}

```
Program Correct8Errors;
Uses CRT, GFUn, SKeyUn;
```

```
Var     TextIn  : BA; {Source text file, picked up from disk }
        TextTD : BA; {Text plus zeros in time domain}
        TextEr  : BA; {Text in time domain plus errors}
        TextRx : BA; {Text plus errors back in frequency domain}
        ErrSpc  : BA; {Completed error spectrum in frequency domain}
        ErrPat  : BA; {Time domain error pattern}
        TextEC : BA; {Error corrected text in time domain}
        TextOu : BA; {Original text after error recovery}

        Ftn             : File;
        i               : Byte;
```
```
Begin
      ClrScr;
      TextColor (10);
      If GFGen (285) Then
      Begin
            FillChar (TextIn, 16, 0);
            Assign (Ftn, 'test.txt');
            Reset (Ftn, 1);
            BlockRead (Ftn, TextIn[16], 255-16);
            Close (Ftn);

            DisplayChar (@TextIn, 255, 1, 3, 13);
            TextColor (10);GotoXY (2, 1); Write ('Original  data');

            IFT (@TextIn, @TextTD);
            DisplayChar (@TextTD, 255, 19, 3, 11);
            TextColor (10);GotoXY (20, 1);Write ('Encoded message');
            Move (TextTD, TextEr, 255);
```

```
{Add eight errors to message}
                TextEr [$46]        := TextEr [$46] XOR $61;
                TextEr [$78]        := TextEr [$78] XOR $84;
                TextEr [$E1]        := TextEr [$E1] XOR $18;
                TextEr [$22]        := TextEr [$22] XOR $AA;
                TextEr [$4F]        := TextEr [$4F] XOR $D2;
                TextEr [$A4]        := TextEr [$A4] XOR $C3;
                TextEr [$CC]        := TextEr [$CC] XOR $30;
                TextEr [$34]        := TextEr [$34] XOR $23;

                DisplayChar (@TextEr, 255, 37, 3, 11);
                TextColor (10);GotoXY (37, 1);Write ('Corrupted message');

                FFT (@TextEr, @TextRx);
                DisplayChar (@TextRx,255, 55, 3, 13);
                TextColor (10);GotoXY (56, 1);Write ('Corrupted data');

                DisplayHex (@TextRX, 15, 12, 22);
                TextColor (10);GotoXY (13, 20);
                Write ('Top row of Error spectrum from Corrupted data');

                RecursiveExt (@TextRx, 8, 254);
                DisplayChar (@TextRx, 255, 1, 26, 13);
                TextColor (10);GotoXY (2, 24);Write ('Error spectrum');

                IFT (@TextRX, @ErrPat);
                DisplayChar (@ErrPat, 255, 19, 26, 11);
                TextColor (10);GotoXY (19, 24);Write ('TD Error Pattern');

                For i := 0 to 254 do TextEC [i] := TextEr [i] XOR ErrPat[i];

                FFT (@TextEC, @TextOu);

                DisplayChar (@TextOu, 255, 37, 26, 13);
                TextColor (10);GotoXY (38, 24);Write ('Corrected Data');
                GotoXY (12, 48);
                TextColor (9);
                Write ('FREQUENCY DOMAIN CODING EXAMPLE WITH 8 ERRORS');
                TextColor (13); GotoXY (10, 45);
                Write ('Magenta = FD');
                TextColor (11); GotoXY (30, 45);
                Write ('Cyan   = TD');
                TextColor (14); GotoXY (50, 45);
                Write ('Yellow  = FD/HEX');
        End;
        Repeat Until KeyPressed;
End.
```

Appendix D

Time domain error correction in three dimensions

{This programming example extends GF(2^4) into three dimensions and performs an error correction on the data.}

```
Program CorrectErrors;
Uses CRT, GFUn;
```

{Three-dimensional data}
```
Var     D0                      : Array [1..15, 1..15, 1..15] Of Byte;
```
{Syndromes and check symbols}
```
        S0, S1, S2, S3, P, Q, R, S : Byte;
```
{Sundry variables}
```
        x, y, z, l, m, n, i, j, k        : Integer;
```

```
Begin
        ClrScr;
        GFGen (19);
```
{Use the polynomial $x^4+x+1=0$}

{Fill data with random values}
```
        For i := 1 to 15 do
                For j := 1 to 15 do
                        For k := 1 to 15 do D0 [i,j,k] := Random (16);
```

{Set check symbols to zero}
```
        D0[14, 15, 15] := 0; {P}
        D0[15, 14, 15] := 0; {Q}
        D0[15, 15, 14] := 0; {R}
        D0[15, 15, 15] := 0; {S}
```

{Calculate syndromes}
```
        S0 := 0; S1 := 0; S2 := 0; S3 := 0;
        For i := 1 to 15 do
        For j := 1 to 15 do
        For k := 1 to 15 do
        Begin
                S0 := S0 XOR D0[i,j,k];
                S1 := S1 XOR GF2Real((AxB(Real2GF (D0[i,j,k]), ApB (1,i))));
                S2 := S2 XOR GF2Real((AxB(Real2GF (D0[i,j,k]), ApB (1,j))));
                S3 := S3 XOR GF2Real((AxB(Real2GF (D0[i,j,k]), ApB (1,k))));
        End;
```

{Convert syndromes to powers}
```
        S0 := Real2GF (S0);
        S1 := Real2GF (S1);
        S2 := Real2GF (S2);
        S3 := Real2GF (S3);
```

{Solve for P, Q, R, and S}
```
        P := AoB (AaB (S0, S1), AaB (14, 15));
        Q := AoB (AaB (S0, S2), AaB (14, 15));
        R := AoB (AaB (S0, S3), AaB (14, 15));
        S := AaB (AaB (P, Q), AaB (S0, R));
```

{Add check symbols to data}
```
        D0[14, 15, 15] := GF2Real (P); {P}
        D0[15, 14, 15] := GF2Real (Q); {Q}
        D0[15, 15, 14] := GF2Real (R); {R}
        D0[15, 15, 15] := GF2Real (S); {S}
```

{Generate a 3 dimensional error location at l, m, n}
```
        l       := 1+Random (15);
        m       := 1+Random (15);
```

```
        n         := 1+Random (15);
```

{Corrupt the message at l, m, n with pattern 1100}
```
        D0 [l, m, n] := D0 [l, m, n] XOR 12;
```

{Calculate syndromes at receiver}
```
        S0 := 0; S1 := 0; S2 := 0; S3 := 0;

        For i := 1 to 15 do
        For j := 1 to 15 do
        For k := 1 to 15 do
        Begin
                S0 := S0 XOR D0[i,j,k];
                S1 := S1 XOR GF2Real((AxB(Real2GF (D0[i,j,k]), ApB (1,i))));
                S2 := S2 XOR GF2Real((AxB(Real2GF (D0[i,j,k]), ApB (1,j))));
                S3 := S3 XOR GF2Real((AxB(Real2GF (D0[i,j,k]), ApB (1,k))));
        End;
```

{Calculate error location from syndromes}
```
        x := AoB (Real2GF (S1), Real2GF (S0));
        y := AoB (Real2GF (S2), Real2GF (S0));
        z := AoB (Real2GF (S3), Real2GF (S0));
```

{Output actual error and located error}
```
        WriteLn ('l = ',l,' m = ',m,' n = ',n);
        WriteLn ('x = ',x,' y = ',y,' z = ',z);
```

{Correct data}
```
        D0 [x,y,z] := D0 [x,y,z] XOR S0;

        Repeat Until KeyPressed;
End.
```

Appendix E

Correcting two-dimensional data in the frequency domain

{This programming example uses frequency domain error correction over two-dimensional data to locate and correct two errors.}

Program CorrectErrors;
Uses CRT, GFUn;

Type D2 = Array [0..6, 0..6] Of Byte;

Var	DataIn	: D2; {**2D input data plus zeros (frequency domain)**}
	DataTD	: D2; {**2D time domain message - no errors**}
	ErrrTD	: D2; {**2D time domain error pattern**}
	DataEr	: D2; {**2D corrupted TD message**}
	DataFE	: D2; {**2D frequency domain message plus errors**}
	ErrSpc	: D2; {**2D reconstructed error spectrum**}
	RErrTD	: D2; {**2D reconstructed TD error pattern**}
	DataOu	: D2; {**2D corrected data out**}

{**Sundry variables**}
　　　　i, j　　　　　　　: Integer;
{**REC feedback elements**}
　　　　Li1, Li2, Lj1, Lj2: Byte;
{**REC registers**}
　　　　Q0, Q1, Q2　　 : Byte;

Begin
　　　　ClrScr;
　　　　GFGen (11);
　　　　Repeat

{**Generate random data**}
　　　　　　　　For i := 0 to 6 do
　　　　　　　　　　　　For j := 0 to 6 do
　　　　　　　　　　　　　　　　DataIn [i,j] := Random (8);
{**Add zeros for a double correcting code**}
　　　　　　　　DataIn [0,0] := 0;
　　　　　　　　DataIn [1,0] := 0;
　　　　　　　　DataIn [2,0] := 0;
　　　　　　　　DataIn [3,0] := 0;

　　　　　　　　DataIn [0,1] := 0;
　　　　　　　　DataIn [0,2] := 0;
　　　　　　　　DataIn [0,3] := 0;
　　　　　　　　DataIn [1,1] := 0;

　　　　　　　　Display2D (@DataIn, 1, 1);

{**Convert FD data to TD message**}
　　　　　　　　IFT2 (@DataIn, @DataTD);
　　　　　　　　Display2D (@DataTD, 1, 9);

{**Generate a 2D error**}
　　　　　　　　FillChar (ErrrTD, SizeOf (ErrrTD), 0);
　　　　　　　　ErrrTD [4,5] := 6;
　　　　　　　　ErrrTD [2,2] := 5;

{**Add error to message**}
　　　　　　　　For i := 0 to 6 do
　　　　　　　　　　　　For j := 0 to 6 do
　　　　　　　　　　　　　　　　DataEr [i,j] := DataTD [i,j] XOR ErrrTD [i,j];

{Produce corrupted frequency domain error}
        FFT2 (@DataEr, @DataFE);
        Display2D (@DataFE, 17);

{Convert FD message + errors to GF elements}
        For i := 0 to 6 do
                For j := 0 to 6 do
                        DataFE[i,j] := Real2GF (DataFE[i,j]);

{Using E00 to E30, find Li1 and Li2}
        Li1 := AoB (AaB (AxB (DataFE [0,0], DataFE [3,0]),
                                AxB (DataFE [1,0], DataFE [2,0])),
                           AaB (AxB (DataFE [1,0], DataFE [1,0]),
                                  AxB (DataFE [0,0], DataFE [2,0])));

        Li2 := AoB (AaB (AxB (DataFE [1,0], DataFE [3,0]),
                                AxB (DataFE [2,0], DataFE [2,0])),
                           AaB (AxB (DataFE [1,0], DataFE [1,0]),
                                  AxB (DataFE [0,0], DataFE [2,0])));

{Reconstruct top row of error specturm}
        Q2 := DataFE [0,0];
        Q1 := DataFE [1,0];
        For i := 0 to 6 do
        Begin
                ErrSpc [i,0] := GF2Real (Q2);
                Q0 := AaB (AxB (Q2, Li2), AxB (Q1, Li1));
                Q2 := Q1;
                Q1 := Q0;
        End;

{Using E00 to E03, find Lj1 and Lj2}
        Lj1 := AoB (AaB (AxB (DataFE [0,0], DataFE [0,3]),
                                AxB (DataFE [0,1], DataFE [0,2])),
                           AaB (AxB (DataFE [0,1], DataFE [0,1]),
                                  AxB (DataFE [0,0], DataFE [0,2])));

        Lj2 := AoB (AaB (AxB (DataFE [0,1], DataFE [0,3]),
                                AxB (DataFE [0,2], DataFE [0,2])),
                           AaB (AxB (DataFE [0,1], DataFE [0,1]),
                                  AxB (DataFE [0,0], DataFE [0,2])));

{Reconstruct first column of error spectrum}
        Q2 := DataFE [0,0];
        Q1 := DataFE [0,1];

        For j := 0 to 6 do
        Begin
                ErrSpc [0,j] := GF2Real (Q2);
                Q0 := AaB (AxB (Q2, Lj2), AxB (Q1, Lj1));
                Q2 := Q1;
                Q1 := Q0;
        End;

{Using Li1 and Li2, E01 and E11, complete second row}
        Li1 := AxB (1, Li1); {Multiply Li1 by α}
        Li2 := AxB (2, Li2); {Multiply Li2 by α^2}

        Q2 := DataFE [0,1];
        Q1 := DataFE [1,1];

```
            For i := 0 to 6 do
            Begin
                        ErrSpc [i,1] := GF2Real (Q2);
                        Q0 := AaB (AxB (Q2, Li2), AxB (Q1, Li1));
                        Q2 := Q1;
                        Q1 := Q0;
            End;
```

{Now that we have two rows, we can complete all the columns}

```
            For i := 1 to 6 do
            Begin
                        Lj1 := AxB (1, Lj1); {Multiply Lj1 by α}
                        Lj2 := AxB (2, Lj2); {Multiply Lj2 by α²}
                        Q2 := Real2GF (ErrSpc [i,0]);
                        Q1 := Real2GF (ErrSpc [i,1]);
                        For j := 0 to 6 do
                        Begin
                                    ErrSpc [i,j] := GF2Real (Q2);
                                    Q0 := AaB (AxB (Q2, Lj2), AxB (Q1, Lj1));
                                    Q2 := Q1;
                                    Q1 := Q0;
                        End;
            End;
```

{Convert the complete error spectrum back to a time domain pattern}

```
            IFT2 (@ErrSpc, @RErrTD);
            Display2D (@ErrSpc, 25, 1);
            Display2D (@RErrTD, 25, 9);
      Until ReadKey = 'e';
End.
```

Appendix F

Example of mixed domain error correction

{This program shows how a combination of time and frequency domain processing can be used to simplify the error coding problem and speed up decoding. The program introduces and subsequently corrects four errors in a time domain message.}

```
Program GFA4;
Uses CRT, GFUn, GFMatUn;

Var     RD, FT, ER                      : Array [0..255] Of Byte;
        S, L, Q                         : Array [0..7] Of Byte;
        i, j                                    : Word;
        ME, ML, MS, MC, MI, M           : MatP;
        ErrPos, ErrPat                  : Array [0..3] Of Byte;
Begin
        ClrScr;
```

{Operate over default field **GF(2^8)** using the polynomial 285}

{Generate random data}
```
        For i := 0 to 254 do RD [i] := Random (256);
```

{For four symbol errors, set first eight symbols to zero}
```
        FillChar (RD [0], 8, 0);
```

{Calculate syndromes}
```
        For j := 0 to 7 do
        Begin
                S[j] := 0;
                For i := 0 to 254 do
                        S[j] := AaB(S[j], AxB(ApB(1, i*j), Real2GF(RD[i])));
                S[j] := GF2Real(S[j]);
        End;
```

{Generate check symbol generator matrix. Since this does not depend on the data, it could be pre-defined and stored in tables.}
```
        M := MakeSMatrix (8);
        For i := 0 to 7 do
                For j := 0 to 7 do
                        M^.P^[i*8+j] := GF2Real(ApB(1, i*j));
```

{Invert matrix to solve}
```
        MI := InvertMatrix (M);
        KillSMatrix (M);
```

{Copy syndromes to a matrix}
```
        MS := MakeRMatrix (1,8);
        Move (S, MS^.P^, 8);
```

{Multiply syndromes by solution matrix}
```
        MC := TimesMatrix (MI, MS);
```

{Copy the eight check symbols to their positions in TD message}
```
        Move (MC^.P^, RD, 8);
```

{Free-up temporary matrix variables}
```
        KillRMatrix (MS);
        KillSMatrix (MI);
        KillRMatrix (MC);
```

{Transmit message and add four errors}
        For i := 0 to 3 do
        Begin
                ErrPos [i] := Random (255);
                ErrPat [i] := Random (255);
                RD [ErrPos [i]] := RD[ErrPos [i]] XOR ErrPat [i];
        End;

{Calculate syndromes at receiver}
        For j := 0 to 7 do
        Begin
                S[j] := 0;
                For i := 0 to 254 do
                        S[j] := AaB(S[j], AxB(ApB(1, i*j), Real2GF(RD[i])));
                S[j] := GF2Real(S[j]);
        End;

{Display non-zero syndromes}
        DisplayDec (@S, 8, 1, 1);

{Either Fourier transform received message to FT or use syndromes since these equal the required spectra, and REC. In this case we'll use the latter since the syndromes have already been calculated in a bid to test for errors.}
{   **FFT (@RD, @FT);**   }
        Move (S, FT, 8);
        RecursiveExt (@FT, 4, 254);

{Construct time domain errors}
        IFT (@FT, @ER);
{Display errors}
        DisplayDec (@ER, 255, 1, 4);

        WriteLn;
        WriteLn;

{Display actual added errors}
        For i := 0 to 3 do
        Begin
                WriteLn ('ErrPos ',i,' = (',ErrPos [i] Mod 16,',',ErrPos [i] DIV
                                16,') Pattern = ',ErrPat[i]);
        End;

{Correct errors}
        For i := 0 to 254 do RD[i] := RD[i] XOR ER[i];
        ReadKey;
End.

Index